Green, Brown, and Probability
& Brownian Motion on the Line

Green, Brown, and Probability
& Brownian Motion on the Line

KAI LAI CHUNG
Stanford University, USA

World Scientific
New Jersey • London • Singapore • Hong Kong

Published by

World Scientific Publishing Co. Pte. Ltd.

P O Box 128, Farrer Road, Singapore 912805

USA office: Suite 1B, 1060 Main Street, River Edge, NJ 07661

UK office: 57 Shelton Street, Covent Garden, London WC2H 9HE

British Library Cataloguing-in-Publication Data
A catalogue record for this book is available from the British Library.

ISBN 981-02-4689-7
ISBN 981-02-4690-0 (pbk)

Printed by FuIsland Offset Printing (S) Pte Ltd, Singapore

PREFACE

In this exposition I hope to illustrate certain basic notions and methods of probability theory by unraveling them in a simple and fruitful case: that of the Brownian Motion Process applied to a pair of famous problems in Electricity. One is the boundary value problem initiated by Green which begins this book, and the other is the equilibrium problem formerly associated with an energy principle, which concludes this book. The first problem is solved by the distribution of the Brownian paths at their first exit from Green's conductor, and the second problem by that of the same paths at their last exit therefrom. To adopt a classic Chinese rhetorical turn: "if one could resurrect and ask him" [若能起而問之] what would Green say?

The book is short enough not to require a summary or abstract. The undecided reader can easily leaf through the hundred generously spaced, well printed pages to see if he/she is "willing and able". To quote Emerson: "'Tis the good reader that makes the good book". In my writing I have taken a generally inquisitive, tutorial stance, interposed some amusing history and personal reminiscences, indulged in a few cavils and slangs, frequently taken the reader into my confidence, with quizzing and teasing. Here are some vista points not to be missed: my inspiration comes from the Eulerian integral over time of the DeMoivre–Laplace–Gauss normal density that yields the Newton–Coulomb potential, exhibited in (2.3). The thread of probabilistic thinking unrolls from the spontaneous splitting of the time-arrow exhibited in (3.4). The great random march starts with the "post-option-temporal-homogeneity" announced in (4.11). From this it is really a short step, barring formalities, to Green's electrical discoveries embodied in (8.9) and (8.11).

Although the contents had been maturing in my mind over years, I did the final writing in four months beginning near the end of the year 1994. Even over such a period of time my mood was subject to fluctuations which must have affected the selection of material as well as its presentation. Suffice it to tell that for each Note preserved in the final version another probably has been discarded, and no claim is made to any uniform standard of relevance or circumspection. I have also been quite casual about the references, not all of which are fully documented or listed. As an innovation I have appended six photocopies of original writings, first pages of books and articles, at the end of the volume, not as museum pieces but as moments of recaptured time.

Portions of the book have been rehearsed in various audiences, in: Mexico (1991), Dalian, China (1991), Lawrence, Kansas (1992), Columbia, Missouri (1992), Auburn, Alabama (1995), Milan, Italy (1995). It is a pleasure to take this opportunity to thank my hostess and hosts on those memorable occasions: Maria Emilia Caballero, Diego Hernandez (deceased), Litze Xu, Tyrone Duncan, Zhongxin Zhao, Olav Kallenberg, Bruno Bassan.

World Scientific Publishing Company is publishing my Chinese book "A New Introduction to Stochastic Processes", (written with the assistance of Rong Wu). Its undertaking to publish this somewhat unconventional "new venture", with the tremendous efficiency and splendid cooperation of its staff, deserves particular recognition and appreciation not only on my part, but also, let us hope, by prospective readers of the World.

> Kai Lai Chung
> July 3, 1995
> Stanford, California

In the new edition I have added the important Addenda and inserted some titbits of new information at various places. These were acquired through correspondence with acquaintances or from welcome strangers. The names of contributors are given without titles, as before. In March 2000 I received a copy of "George Green" (The Athlone Press, 1995) from the authoress D. M. Cannell, accompanied by a letter. She answered my "speculation as to whether Green and Brown ever met", saying that she "would be prepared to say (if not to publish!) that they never did". I then wrote to ask her if there exists a photo or painting of George Green, since he does not appear in the twenty plates reproduced in her book, including Green's family mill which has been restored. Unfortunately she died shortly after and it was Lawrie Challis who wrote to say "no". Such a pity. I am indebted to Robert E. O'Malley Jr. who mentioned my book to Green's biographer.

> March 17, 2001

CONTENTS

Part II. Brownian Motion on the Line

Part III. Stopped Feynman–Kac Functional

Part I.

Green, Brown, and Probability

SOME NOTATION
AND TERMINOLOGY

R^d is Euclidean space of dimension d.

$R_0 = [0, \infty)$.

$\mathcal{B}$ is the class of Borel sets in R^d, called the Borel tribe.

$\mathcal{B}_b$ is the class of bounded Borel measurable functions.

$\mathcal{B}_c$ is the class of Borel measurable functions which vanishes outside some unspecified compact set; it is also said to have compact support.

$C(D)$ is the class of continuous function on the set D; when (D) is omitted (R^d) is meant.

C_b, C_c are defined like $\mathcal{B}_b$, $\mathcal{B}_c$ and usually come with (D) as in $C(D)$.

$C^2(D)$ is the subclass of C which are twice continuously differentiable in the open set D; you can guess what C^n and C^∞ mean even with the subscript c adjoined, but watch out if C_b^2 is used it could mean different things.

m is the Borel–Lebesgue measure in any R^d, dm is used as the differential but dx or dy is also used.

ω is used for the often omitted sample point in the sample space Ω.

3

BMP is about the only acronym used here and reads Brownian Motion Process, which is also referred to without the acronym, and often without the word Process; the word Movement is also used instead of Motion.

"Almost surely" or "a.s." means "with probability one"; more particularly, "P^x-a.s." restricts the probability to P^x.

1. GREEN'S IDEAS

In order to tell the story a little in its historical perspective, I shall begin with George Green's original ideas, published in 1828.[1]

A formula is customarily taught in a second year calculus course, known as Green's Second Identity:

$$\int_D (u\Delta v - v\Delta u)dm = \int_{\partial D} \left(u\frac{\partial v}{\partial n} - v\frac{\partial u}{\partial n} \right) d\sigma . \qquad (1.1)$$

Here D is a "region" in R^3, ∂D its bounding surface, dm (usually written as dx) is the volume element, $d\sigma$ is the area element, u and v are two functions, Δ (sometimes written as ∇^2) is the Laplacian, and $\frac{\partial}{\partial n}$ is the outward normal derivative. In an elementary course these terms are introduced casually; for instance what exactly is a "region"? We shall not bother with tedious formalities until later. As a matter of fact, a rigorous demonstration of the identity under appropriate conditions cannot be attempted in a sophomore or even a graduate class.

Let us begin with D. We assume it to be an open, nonempty (how stupid to have to say that!) and connected set, and call such a set a "domain" from here on, abandoning the often ill-defined term "region".

A function v is called "harmonic" in D iff $v \in C^2(D)$, and satisfies the Laplace equation in D:

$$\Delta v = 0, \qquad \Delta = \sum_{i=1}^{3} \frac{\partial^2}{\partial x_i^2} . \qquad (1.2)$$

Without further ado, we will apply (1.1) with an arbitrary harmonic function v and a very special function u. The latter carries a "parameter" and will be denoted by $u_x(\cdot)$ or $u(x, \cdot)$:

$$u_x(y) \equiv u(x, y) \equiv \frac{1}{|x - y|} . \qquad (1.3)$$

You may recognize this as Newton's gravitational potential, as well as Coulomb's electrostatic potential. Apart from a numerical constant, of the utmost importance in physics (and consequently for the real world), but quite ignorable in mathematics, $u(x, y)$ represents the potential induced by a mass or charge placed at the point x and evaluated at the point y. For each x, we have (Δ_y is Δ with y as variable)

$$\Delta_y u_x(y) = 0 \qquad \text{if } y \neq x ; \qquad (1.4)$$

namely u_x is harmonic in $R^3 \backslash \{x\}$.

You should verify (1.4) as an exercise although it is a dull one. You may have already done so in your calculus course, and now you will find out what it is all about. Owing to the little exception indicated in (1.4), Green's identity (1.1) is NOT applicable with u_x as the u there when the domain D contains the point x, as we do assume. We can get around the difficulty by surrounding x with a small ball $B = B(x, r)$, such that $\bar{B} \subset D$, and apply (1.1) to the domain $D \backslash \bar{B}$.[2] By the way, from now on we will assume D to be bounded. Since $\Delta_y v = \Delta_y u_x = 0$ in $D \backslash \bar{B}$, and $\partial(D \backslash \bar{B}) = (\partial D) \cup (\partial B)$, (1.1) reduces to

$$0 = \int_{\partial D} \left(u_x \frac{\partial v}{\partial n} - v \frac{\partial u_x}{\partial n} \right) d\sigma + \int_{\partial B} \left(\frac{1}{r} \frac{\partial v}{\partial n} - \frac{1}{r^2} v \right) d\sigma \qquad (1.5)$$

because on ∂B, we have $u = \frac{1}{r}$ and $\frac{\partial u}{\partial n} = -\frac{1}{r^2}$. The last integral can be evaluated by the mean-value theorem (of integrals!) as

$$\left(\frac{1}{r} \frac{\partial v}{\partial n}(y') - \frac{1}{r^2} v(y'') \right) 4\pi r^2$$

where y' and y'' are two points on ∂B, and $4\pi r^2$ is the area of ∂B. [Not many American students know or care to know the area of a sphere or the volume of a ball; but I hope you do.] Since v is harmonic in D, both it and $\frac{\partial v}{\partial n}$ are bounded on ∂B as r decreases to zero, while $v(y'')$ converges to $v(x)$. Thus when we shrink B to $\{x\}$ we obtain from (1.5):

$$-4\pi v(x) = \int_{\partial D} \left(\frac{\partial u_x}{\partial n} v - u_x \frac{\partial v}{\partial n} \right) d\sigma, \qquad x \in D . \qquad (1.6)$$

This relation holds for any harmonic v and any x in D. As Green observed, it *represents* v by its boundary values together with those of $\frac{\partial v}{\partial n}$. But the latter is a nuisance and must be got rid of. How? Here is his bright idea. Apply (1.1) once more with the same v but another truly harmonic h_x for u, so that this time there is no trouble and (1.1) yields at once

$$0 = \int_{\partial D} \left(\frac{\partial h_x}{\partial n} v - h_x \frac{\partial v}{\partial n} \right) d\sigma . \qquad (1.7)$$

Now subtract (1.7) from (1.6):

$$-4\pi v(x) = \int_{\partial D} \left(\frac{\partial (u_x - h_x)}{\partial n} v - (u_x - h_x) \frac{\partial v}{\partial n} \right) d\sigma . \qquad (1.8)$$

In this formula $h_x(\cdot)$ is any function harmonic in D. If we can choose it so that it be equal to $u_x(\cdot)$ on ∂D, then the unwelcome term involving $\frac{\partial v}{\partial n}$ in (1.8) will be knocked off, and the result will be a representation of v by its boundary values, as desired:

$$v(x) = \int\limits_{\partial D} \frac{-1}{4\pi} \frac{\partial}{\partial n} (u_x - h_x) \cdot v d\sigma . \tag{1.9}$$

The problem of finding such an h_x became known as the "first boundary value problem" of Dirichlet's. But for Green it was "no problem" (*pas de problème*) at all. Why, the solution is given by NATURE: h_x is nothing but the potential of the charge induced on the grounded surface by a unit charge at x; provided that you understand this lingo of electromagnetism! Here I should like to quote from Oliver Dimon Kellogg ([pp. 237–238]):

> "We have here an excellent example of the value and danger of intuitional reasoning. On the credit side is the fact that it led Green to a series of important discoveries since well established. On the debit side is its unreliability, for there are, in fact, regions for which Green's function does *not* exist."

We shall solve the boundary value problem by probabilistic method. For the moment suppose the requisite h_x has been found, and suppose that it has a continuous normal derivative on ∂D as appearing in (1.9) — something Green probably never worried about but Kellogg did work on. Then it is clear that the problem is equivalent to the determination or construction of the function

$$(x, y) \rightarrow u_x(y) - h_x(y) . \tag{1.10}$$

This function of two variables is now called Green's function. apart from a numerical factor which still varies from author to author. Many books with it in the title have been written.

2. PROBABILITY
AND POTENTIAL

In the scenario of the preceding discussion, there is only space and no time. But it is generally perceived that TIME is the mysterious element of our universe. Consider the function of three variables which should be familiar if you have studied any probability or statistics:

$$p(t; x, y) = \frac{1}{(2\pi t)^{d/2}} e^{-\frac{|x-y|^2}{2t}} \tag{2.1}$$

where $t \in (0, \infty)$, $x \in R^d$, $y \in R^d$, $d \geq 1$. For fixed x, $p(t; x, \cdot)$ is the normal density function with mean x and variance t. If we regard the parameter t as time, it becomes natural to integrate "over time". For $d = 1$ and $d = 2$, the result is the disappointing $+\infty$. For $d \geq 3$, however, we get the fascinating

$$\int_0^\infty p(t; x, y) dt = \frac{\Gamma\left(\frac{d}{2} - 1\right)}{2\pi^{d/2}} \frac{1}{|x-y|^{d-2}} . \tag{2.2}$$

You should carry out the "definite integral" above without recourse to a table or handbook, thereby learning Euler's integral for his gamma function Γ. You should also find out the value of $\Gamma(\frac{1}{2})$ which will yield $\frac{1}{2\pi|x-y|}$ for the right-hand member of (2.2) when $d = 3$. From now on we take $d = 3$, although for all $d \geq 3$ the results are quite similar except

9

for certain constants depending on d. The case $d = 1$ can be treated by easier methods (what is a harmonic function in R^1?). The case $d = 2$ is unfortunately more difficult and must be left out in spite of its appeal and import.

Our first recognition is that integrating $p(t; x, y)$ over all time t leads us to the Newtonian potential (1.3) apart from a numerical factor. Let us introduce a new notation to absorb the latter:

$$\int_0^\infty p(t; x, y)dt = g(x, y) = \frac{1}{2\pi}u(x, y) = \frac{1}{2\pi|x - y|} \, . \qquad (2.3)$$

At the same time, we introduce an operator G as follows: for suitable functions f:

$$Gf(x) = \int g(x, y)f(y)dy = \frac{1}{2\pi}\int \frac{f(y)}{|x - y|}dy \, , \qquad (2.4)$$

where the integral is taken over all R^3, as also in similar cases below. We are using f as a dummy variable in (2.4) in the sense that G maps f into Gf. It is usually sufficient to use a class of functions (called "test functions") to determine the operator G; for example the class $C_c(R^3)$. But if you are not familiar with such doings you should do two exercises in this connection:

(i) if $f \in \mathcal{B}_c$ then $Gf \in C_b$;
(ii) if $f \equiv 1$ then $G1 = \infty$; hence the class C_b is not suitable for (2.4)!

Having introduced G we might as well define the operator P_t associated with $p(t; \cdot, \cdot)$:

$$P_t f(x) = \int p(t; x, y)f(y)dy \, . \qquad (2.5)$$

In view of (2.3), we have then symbolically:

$$Gf = \int_0^\infty P_t f \, dt \quad \text{or} \quad G = \int_0^\infty P_t \, dt . \qquad (2.6)$$

Let me take this opportunity to review an indispensable tool of all analysis: the theorem of Fubini and Tonelli. For a given f, the validity of (2.6) amounts to reversing the order of two integrations in

$$Gf(x) = \int_{R^3} \left(\int_0^\infty p(t; x, y) \, dt \right) f(y) \, dy$$

$$= \int_0^\infty \left(\int_{R^3} p(t; x, y) f(y) \, dy \right) dt = \int_0^\infty P_t f(x) \, dt . \tag{2.7}$$

If $f \geq 0$, then according to Tonelli's supplement to his teacher Fubini's theorem, such a change of order of integrations is always OK, whether the result of the double integration be finite or infinite. Thus we have

$$G|f|(x) = \frac{1}{2\pi} \int_{R^3} \frac{|f|(y)}{|x - y|} \, dy = \int_0^\infty P_t |f|(x) \, dt \leq +\infty . \qquad (2.8)$$

Now if the result above is "$< \infty$", then Fubini's theorem asserts that we can replace $|f|$ everywhere in (2.8) by f and the equations will also hold. This is the kind of application we shall make again and again, in diverse similar circumstances. Allow me to call attention to a common *faux-pas*: we must check $\int_0^\infty P_t |f|(x) \, dt < \infty$ in the step above, not $\int_0^\infty |P_t f|(x) \, dt < \infty$! This sort of mistake has been made by practising mathematicians in a hurry, beware.

The relation (2.3) furnishes a vital link between two big things. At one end we see the Newton–Coulomb potential; at the other end the normal density that is also known as the DeMoivre–Laplace–Gauss distribution, (alias the "error function"). Can the linkage be a mere accident, or does it portend something of innate significance?[3]

I do not know if the clue hidden in (2.3) bridging the random and the deterministic had been noticed by earlier mathematicians and physicists (they were the same group formerly). Oddly enough, there was an intervening development, from botany.

Robert Brown was a contemporary of George Green, though he was a Scot, not an Englishman. He was a noted botanist who made his microscopic observation of plant-pollen movements in 1827, published in 1828, the same year of Green's Essay. (Did the two ever meet?) Tiny particles suspended in liquid move about in a highly irregular fashion. Naturally their paths have got to be continuous, but they were seen apparently to be so irregular that they were believed to be non-differentiable by the physicist Jean Perrin, who had heard about Weierstrass's discovery of such pathological functions. [I learned this first around 1937 from Titchmarsh's Theory of Functions.] But it was Albert Einstein[4] who first used a probability model to study the random movements of those pollen particles that became known as Brownian motion.[5]

Einstein's original papers on the subject are now available in English. Here is a slightly edited version of the relevant portion. (see [Einstein; pp. 14–15]). In a short time interval τ the relative proportion of particles which experience a displacement between δ and $\delta + d\delta$ can be expressed by probability law $\varphi(\delta)$ such that

$$\int\limits_{-\infty}^{\infty} \varphi(\delta)d\delta = 1; \qquad \int\limits_{-\infty}^{\infty} \delta\varphi(\delta)d\delta = 0;$$

$$\frac{1}{2}\int\limits_{-\infty}^{\infty} \delta^2\varphi(\delta)d\delta = D\tau , \qquad (2.9)$$

where D is the coefficient of diffusion. Put $f(x,t)$ to be the number of particles per unit volume at location x and time t. Then on the one hand we have

$$f(x,t+\tau) = f(x,t) + \frac{\partial f}{\partial t}\tau ;$$

and on the other hand we have

$$f(x,t+\tau) = \int\limits_{-\infty}^{\infty} f(x+\delta,t)\varphi(\delta)d\delta$$

$$= \int\limits_{-\infty}^{\infty} \left[f(x,t) + \frac{\partial f}{\partial x}\delta + \frac{1}{2}\frac{\partial^2 f}{\partial x^2}\delta^2 \right] \varphi(\delta)d\delta$$

in which the higher infinitesimals have been ignored. Hence by confronting the two approximations, one with respect to time and the other with respect to (randomized) displacements, he obtained the partial differential equation:

$$\frac{\partial f}{\partial t} = D\frac{\partial^2 f}{\partial x^2} \qquad (2.10)$$

from which he concluded that

$$f(x,t) = \frac{n}{\sqrt{4\pi Dt}}e^{-\frac{x^2}{4Dt}} \qquad (2.11)$$

where n is the total number of particles.

In his argument it was assumed that the movements of the single particles are mutually independent. This is standard in statistical mechanics, formerly called "gas theory". His basic assumptions for the movement of the particles will be quoted and discussed later in Sec. 5. Let us also observe the crucial assumption contained in the third equation in (2.9) that makes (2.10) possible. We shall return to this with hindsight. For the moment we note that if $D = \frac{1}{2}$ then the $f(x,t)$ in (2.11) reduces to the $p(t; 0, x)$ in (2.1) with $d = 1$. It sufficed for Einstein to treat one of the three coordinates, of course. The extension of (2.10) to R^3 is known as the heat equation:

$$\frac{\partial f}{\partial t} = \frac{1}{2}\Delta f \ . \tag{2.12}$$

It is a good exercise to verify that for each x, $p(t; x, \cdot)$ satisfies this equation. Remembering Eqs. (1.4) and (2.3), it should arouse our curiosity if we can use the (2.12) for $p(t; x, \cdot)$ to confirm (1.4)? Indeed we can, as follows:

$$\Delta_y u(x, y) = \Delta_y \int\limits_0^\infty p(t; x, y)dt$$

$$= \int\limits_0^\infty \Delta_y p(t; x, y)dt = \int\limits_0^\infty 2\frac{\partial}{\partial t}p(t; x, y)dt$$

$$= 2p(\infty; x, y) - 2p(0+; x, y) = 0 - 0 = 0 \tag{2.13}$$

provided $x \neq y$! This may be heuristic but since it comes out OK, it must be justifiable. It is most satisfying to see that we need $x \neq y$ to justify $\lim\limits_{t\to 0} p(t; x, y) = 0$ in the final step above.

3. PROCESS

We are going to set up a probabilistic model known as the Brownian Motion Process, to be abbreviated to BMP in later references. A stochastic process is a parametrized collection of random variables on a probability space $(\Omega, \mathcal{F}, P)$. The parameter will be called "time" and denoted by t. In the case in question, $t \in [0, \infty) \equiv R_0$ so that time has a beginning but no end. The process is then denoted by $\{X_t\}$ with the range of t understood. For each t, the random variable X_t is the function $\omega \to X_t(\omega)$ from Ω to a state space which will be R^d, usually with $d = 3$ as in previous discussions.

It is my intention to formulate the hypotheses for the BMP in several stages, as they are needed in the gradual unraveling of the basic properties of the process. The first hypothesis is based on Einstein's calculations.

For each $t > 0$,

$$X_t - X_0 \quad \text{has the probability density } p(t; 0, \cdot) \,. \quad (3.1)$$

The distribution of X_0, (viz. the initial distribution of the process) is not specified. If it is the point mass ε_x, namely if $X_0 = x$, then X_t has the density $p(t; x, \cdot)$. Probabilities and mathematical expectations under the assumption $X_0 = x$ will be denoted by P^x and E^x respectively. Thus $P^x\{X_t \in A\} = \int_A p(t; x, y)dy$.

15

Once we have a process we can make up new random variables. For each ω, consider the time integral:

$$\int_0^\infty f(X_t(\omega))dt \ , \qquad \omega \in \Omega \ . \tag{3.2}$$

For a suitable f this will exist, and the resulting function of ω will be a random variable, namely $\mathcal{F}$-measurable. But does it signify anything? When f is the indicator function 1_B of a Borel set B, the integral $\int_0^\infty 1_B(X_t)dt$ represents the total "occupation time" of B, an interesting quantity in certain problems. Let us calculate the mean of the random variable in (3.2). Using Fubini–Tonelli, we obtain

$$E^x\left\{\int_0^\infty f(X_t)dt\right\} = \int_0^\infty E^x\{f(X_t)\}dt$$

$$= \int_0^\infty P_t f(x)dt = Gf(x) \ . \tag{3.3}$$

Thus the operator G introduced in Sec. 2 can be endowed with a probability meaning. Next, take any time T and split the integral in (3.2) as follows:

$$\int_0^\infty f(X_t)dt = \int_0^T f(X_t)dt + \int_T^\infty f(X_t)dt$$

$$= \int_0^T f(X_t)dt + \int_0^\infty f(X_{T+t})dt \ . \tag{3.4}$$

To exploit this seemingly innocuous relation to great advantage, we will coin a new name and a new notation. "Notation can be incredibly important!" (see [R]). Denoting X_{T+t} by X_t^T, we re-invent a new process $\{X_t^T\}$, $t \in R_0$, to be called the "post-T process". For $T = 0$, it is the original $\{X_t\}$. The distribution of the initial random variable $X_0^T \equiv X_T$ is determined by that of X_0, by Hypothesis (3.1), as follows:

$$P^x\{X_0^T \in A\} = P_T(x, A) = \int_A p(T; x, y)dy \qquad (3.5)$$

for any Borel set A. More conveniently, we have for suitable f, for example $f \in C_b$:

$$E^x\{f(X_0^T)\} = P_T f(x) . \qquad (3.6)$$

From this we derive that if X_0 has the distribution μ, then the distribution of X_T will be $\int \mu(dx) P_T(x, \cdot)$.

Now we state the second fundamental hypothesis for $\{X_t\}$.

For each $T \geq 0$, and $t > 0$, $\quad X_t^T - X_0^T$

has the probability density $p(t; 0, \cdot)$. $\qquad (3.7)$

This includes (3.1) and the "temporal homogeneity" of the process, namely for each $T > 0$, the post-T process obeys the same probability law as the original process, except in its initial distribution. Thus the hypothesis is a natural one for botany or physics, in which the results of observation or experimentation should not depend on the specific time when it is carried out, "provided that the temperature, pressure, humidity, ... , remain constant".

Thus, the distribution of $\int_0^\infty f(X_t)dt$ under "$X_0 = y$" is the same as that of $\int_0^\infty f(X_t^T)dt$ under "$X_0^T = y$". Let $X_0 = x$, then by (3.5), the distribution of X_0^T is $P_T(x, \cdot)$. By (3.7), the distribution of $\int_0^\infty f(X_t^T)dt$ must be the same as that of $\int_0^\infty f(X_t)dt$ when X_0 has the distribution $P_T(x, \cdot)$. Therefore we have

$$E^x \left\{ \int_0^\infty f(X_t^T)dt \right\} = \int_{R^3} P_T(x, dy) E^y \left\{ \int_0^\infty f(X_t)dt \right\} . \quad (3.8)$$

In view of (3.3), the right member in (3.8) may be written as

$$P_T G f(x) = E^x \{ G f(X_T) \} .$$

Pushing symbolism further, we can record (3.8) using only $\{X_t\}$ as follows:

$$E^x \left\{ \int_0^\infty f(X_{T+t})dt \right\} = E^x \left\{ E^{X_T} \left[\int_0^\infty f(X_t)dt \right] \right\} . \quad (3.9)$$

This is the form of the result to be found in standard texts. It is an abuse of notation to be tolerated: when $T = t$ the two appearances of X_t denote quite different things!

We now return to (3.4) and take E^x to obtain

$$G f(x) = E^x \left\{ \int_0^T f(X_t)dt \right\} + P_T G f(x) . \quad (3.10)$$

Frankly, so far the result can also be derived from a little operator theory, without the intervention of the process $\{X_t\}$.

For we have $P_{T+t} = P_T P_t$, and consequently

$$G = \int\limits_0^\infty P_t dt = \int\limits_0^T P_t dt + \int\limits_0^\infty P_T P_t dt$$

$$= \int\limits_0^T P_t dt + P_T \int\limits_0^\infty P_t dt = \int\limits_0^T P_t dt + P_T G \; ,$$

$$(3.11)$$

which is (3.10) in a symbolic form. So, what's the big deal? It is the next step that will separate the process $\{X_t\}$ from its transitional operator $\{P_t\}$, "the men from the boys", and show the true worth of probabilistic methods beyond the reach of mere operators. To emphasize this I will begin a new section.

4. RANDOM TIME

A stochastic process is much more than its transition semigroup. The random variables $\{X_t\}$ generate other random variables that have no counter parts in the other theory. To recognize this it is necessary to describe the process in a more formal way as a function of two variables:

$$(\omega, t) \to X(\omega, t), \qquad \omega \in \Omega, \quad t \in R_0 ; \qquad (4.1)$$

where ω is the dummy variable in the "sample space" Ω, while t is the time as before . To each sample ω from Ω corresponds the function $t \to X(\omega, t)$, namely the "sample function" $X(\omega, \cdot)$ from R_0 to R^3. The latter is also called a path or trajectory. Thus, instead of viewing the process as a collection of random variables indicated by the notation $\{X_t\}$ indexed by t and with ω omitted, we can view it as a universe or collection of sample functions $\{X(\omega, \cdot)\}$ indexed by ω and with t indicated by a dot; or more explicitly as in (4.1) with $X_t(\omega) = X(\omega, t)$.

Now when one has two variables x and y as in planar geometry one often needs to substitute a function, say φ, of x for y to get $(x, \varphi(x))$ which represents a curve: $x \to \varphi(x)$. So we need to substitute a function T of ω for t to get from (ω, t) the random time $(\omega, T(\omega))$ or $\omega \to T(\omega)$.

Let us introduce at once two plenipotent random times which will be our main topics in the sequel. For a Borel set A, we define:

$$T_A(\omega) = \inf\{0 < t < \infty : \quad X(\omega, t) \in A\}\,, \qquad (4.2)$$

$$L_A(\omega) = \sup\{0 < t < \infty : \quad X(\omega, t) \in A\}\,. \qquad (4.3)$$

Of course the time-sets between the braces above may be empty, in which case the standard convention applies:

$$\inf \emptyset = +\infty, \qquad \sup \emptyset = 0 \,.$$

Obviously T_A is the "first entrance time" in A, but owing to the tricky infimum in its definition, it may be better to call it the "hitting time" of A. As for L_A, the natural name is the " last exit time" from A, but I have also used the trickier "quitting time" to match the hitting time.

The two random times T_A and L_A have fundamentally different characters. The former is often called a "stopping time", but is better called "optional". To explain optionality it is necessary to bring forth the Borel tribe (σ-field) generated by the random variables of the process. For each t, we denote by $\mathcal{F}_t$ the Borel tribe generated by $\{X_s,\ 0 \le s \le t\}$; for technical reasons we may have to augment this by "null sets", a tedious affair not fit for casual discussion. Then a random variable $\omega \to T(\omega)$ is called optional iff for each t we have $\{T < t\} \in \mathcal{F}_t$. This is described by saying that we can determine whether the random T occurs before t or not by our information on the process up to time t, nothing after t. Even more intuitively, we say that we can choose (opt for) T without clairvoyance. This is actually the language borrowed from a gambling system which inspired J. L. Doob to study such random times in his martingale theory, circa 1940. Apparently the physicists had used the term "first passage time" for T_A, but they did not bother to define mathematical terms since they knew what they were talking about. Now let us take a nice set A and see if we can check that

$$\{T_A < t\} \in \mathcal{F}_t \ . \tag{4.4}$$

This is an instance of the clash between intuition and logic! The Borel tribes are rigidly defined purely mathematical notions, they do not submit facilely to intuitive handling. Even for a simple case as when A is the complement of an open ball, (4.4) is not true without a rather strong assumption on the process $\{X_t\}$ which will be stated shortly. Nevertheless, G. A. Hunt has proved that (4.4) is true for a class of sets A including all Borel sets, even the analytic or Souslin sets (the augmentation of $\mathcal{F}_t$ mentioned above is needed), for a large class of stochastic processes now named after him: Hunt process. It is time to state the next fundamental hypothesis for our BMP.

For each $\omega \in \Omega$, the sample function $X(\omega, \cdot)$

is a continuous function on R_0 to R^3 . $\tag{4.5}$

This is the continuity of the Brownian paths taken for granted by Perrin and others (I do not know if Brown ever asserted it). It was first proved by Norbert Wiener [D] and it depends heavily on the normal distribution in (2.1) as well as the independence assumed by Einstein in his investigation (which is not yet incorporated in the present unraveling). His proof was difficult, as many first proofs were. Paul Lévy who regretted that he missed being first,[6] gave an easier proof which can be read in my recent book [F]. A proof derived from more general considerations can be found in [L]. Now let us see how to show (4.4). We have

$$\{T_A < t\} = \{\exists s \in (0,t): \quad X_s \in A\} \ . \tag{4.6}$$

Warning: this too must be proved. If we change the "<" on the left side to "≤", and simultaneously the "$(0,t)$" to

"$(0, t]$" on the right side, the result becomes false (for an open A)! Now isn't it "intuitively clear" that the right-hand set in (4.6) belongs to $\mathcal{F}_t$? We need to be formal and write this set *logically* as $\bigcup_{s \in (0,t)} \{X_s \in A\}$. Of course for each s in $(0, t)$, the set $\{X_s \in A\}$ belongs to $\mathcal{F}_t$ but the union is over an uncountable collection and nothing in the definition of the Borel tribe $\mathcal{F}_t$ says such a union must be in it! Tribes are creatures of countability. So let us take a countable subset of $(0, t)$, for example the rationals denoted by $\mathbb{Q}$. Can we show that

$$\bigcup_{s \in (0,t)} \{X_s \in A\} = \bigcup_{s \in (0,t) \cap \mathbb{Q}} \{X_s \in A\} ? \qquad (4.7)$$

Here some kind of continuity of X_s in s will help, but we must still subject the set A to a topological condition in order to utilize it. If A is an open set and $X(s)$ is continuous in s, then $X_s \in A$ implies $X_r \in A$ for rational r sufficiently close to s; it follows that (4.7) is true. The argument shows that one-sided continuity of $s \to X_s$ is sufficient. Indeed, it is sufficient that the sample functions $X(\omega, \cdot)$ be all "separable relative to Q", where Q may be any countable set that is dense in R_0. This "separability" is a mini kind of continuity property. According to an ancient theorem of Doob's, any stochastic process has a "version" in which all sample functions have this property.[7] To understand the meaning and use of separability, it is better to read my exposition in [M; II.4] for Markov chains because there it is truly indispensable and used "at every step".

Having proved the optionality of T_A for an open set A, we will take a closed set A and approximate it by its open neighborhoods, as follows. Put $B_n = \{x : d(x, A) < \frac{1}{n}\}$ where "d" denotes distance; then B_n is open, $B_n \supset \bar{B}_{n+1}$, $\bigcap_{n=1}^{\infty} B_n = \bigcap_{n=1}^{\infty} \bar{B}_n = A$. Using the continuity hypothesis in

(4.5), we should have

$$\lim_n T_{B_n} = T_A \qquad (4.8)$$

where the limit is approached strictly increasingly. It then follows that for each $t > 0$:

$$\{T_A \leq t\} = \bigcap_{n=1}^{\infty} \{T_{B_n} < t\} \in \mathcal{F}_t , \qquad (4.9)$$

and we have proved a little more than needed for the optionality of T_A. In these matters hairs must be split: it makes a difference whether we have "$< t$" or "$\leq t$", but here (4.9) is more than (4.4) (why?). I have written down the basic idea of the argument above in order to warn the unsuspecting reader of a *trap*! It is a big mistake or a small mistake according to each one's taste, and I will leave it to you to find it out yourself. Here is a hint: let the path start from A to see what can go wrong with (4.8). The author of a book on Brownian motion, a respectable mathematician of my acquaintance, overlooked this pitfall although it was forewarned in my book [L; p. 72]. There are more than one way to salvage the proof, either by a detour as done in [L; p. 71] or by a more direct method told to me by Giorgio Letta when I lectured on it in Pisa, 1989, which was recorded in [F, p. 19].

Before proceeding further let us explain some notation to be used. For any function $\omega \rightarrow T(\omega)$ we can define the function $\omega \rightarrow X(\omega, T(\omega))$; this is usually denoted by X_T. Similarly for any $t \in R_0$, the function $\omega \rightarrow X(\omega, T(\omega) + t)$ will be denoted by X_{T+t}. But if for some ω, $T(\omega) = +\infty$ as allowed in (4.2), then since $X(\omega, +\infty)$ is undefined we cannot define $X(\omega, T(\omega))$ for that ω. Occasionally it is feasible to extend the definition of $X(\omega, t)$ to $t = +\infty$; but unless this has been done we must restrict X_{T+t} to the set $\{T < \infty\}$. We must also assume that the function $(\omega, t) \to X(\omega, t)$ is jointly measurable $\mathcal{F} \times \mathcal{B}$, where $\mathcal{B}$ is the classic Borel tribe on R_0, in order to assert that X_{T+t} is a random variable, namely $\mathcal{F}$-measurable.

After these preliminaries we are now in the position of extending our previous post-T process from a constant T to a finite optional time T:

$$X_t^T = X_{T+t}, \qquad t \in R_0 . \tag{4.10}$$

Since a constant time is optional (prove it!) this definition includes the previous one in Sec. 3. Observe that when T is optional, so is $T+t$ for each $t > 0$. This is an important remark though we may not see its use below. Now we announce our next, truly revolutionary hypothesis, to be called "post-option homogeneity".

> For each finite optional time T, the post-T process $\{X_t^T\}$ obeys the same probability transition law as the original process $\{X_t\}$. (4.11)

This means, by referring to (3.7), that for each $t > 0$, $X_t^T - X_0^T$ has the probability density $p(t; 0, \cdot)$. Needless to say, the hypothesis (4.5) implies that all sample functions $t \to X(\omega, T(\omega)+t)$ of the post-T process are continuous in R_0. Finally, what happens to (3.5) and (3.6)? P_T is not defined for a random T, yet. We now *define* it through $X_0^T = X_T$, as follows:

$$P_T(x, A) = P^x \{ X_T \in A \} ; \qquad P_T f(x) = E^x \{ f(X_T) \} . \tag{4.12}$$

This notation is invented here for the occasion, it is not "standard" or "habitual", but it will be seen to be most appropriate below.

Under the hypothesis (4.11) the developments in Sec. 3 for a constant time T can be carried over to a finite optional time T, without change, beginning with (3.4) and ending with (3.9) and (3.10). A backward glance at the previous alternative method shown in (3.11) makes it abundantly clear why semigroup is now helpless because a random time T is simply not in its structure.

5. MARKOV PROPERTY

Let D be a bounded domain in R^3, $\bar{D}$ its closure, ∂D its boundary, D^c its complement. The boundary is defined logically by $\bar{D} \cap \overline{D^c}$. It has already appeared in Green's formula (1.1), but in general it can be a tricky set, not necessarily of lower dimension than D, even less to have a surface area, as undoubtedly Green took for granted for his conductors. All we know is that ∂D is a compact set. Since D^c is a closed set, its hitting time T_{D^c} is optional as discussed in Sec. 4. We will call it the "first exit time from D" and denote it by S_D:

$$S_D = \inf\{0 < t < \infty; \quad X_t \in D^c\}. \tag{5.1}$$

We will now prove that almost surely $S_D < \infty$. To do so we need to strengthen the hypothesis (3.7) by bringing in the "past" with respect to the chosen time T, as well as the "future" that is already embodied in the post-T process. For each constant T, the Borel tribe $\mathcal{F}_T$ generated by the collection of random variables $\{X_t, \ 0 \le t \le T\}$ will be called the pre-T tribe; the Borel tribe generated by $\{X_t, \ T \le t < \infty\}$ will be denoted by $\mathcal{F}'_T$ and called the post-T tribe. The "present" X_T belongs to both tribes by this definition; so be it. The post-T tribe is of course just the tribe generated by the post-T process. We could have also defined a pre-T process, but this will not be necessary yet. The single random variable X_T generates a Borel tribe which will be denoted by the same symbol.

You should review the notion of conditional probability and expectation relative to a Borel tribe. The following hypothesis is the formal statement of the "Markov property".

Conditioned on X_T, the tribes $\mathcal{F}_T$ and $\mathcal{F}'_T$ are independent. $\qquad$ (5.2)

This means: for any $\Lambda_1 \in \mathcal{F}_T$ and $\Lambda_2 \in \mathcal{F}'_T$ we have

$$P\{\Lambda_1 \cap \Lambda_2 | X_T\} = P\{\Lambda_1 | X_T\} P\{\Lambda_2 | X_T\} . \qquad (5.3)$$

In this form it looks neat (symmetric!) but it is not very convenient for use. A more handy form is the following: for any $\Lambda \in \mathcal{F}'_T$,

$$P\{\Lambda | \mathcal{F}_T\} = P\{\Lambda | X_T\} . \qquad (5.4)$$

This is often described by words like: "the past has no after-effect on the future when we know the present." But beware of such non-technical presentation (sometimes required in funding applications because bureaucrats can't read mathematics). Big mistakes have been made through misunderstanding the exact meaning of the words "when the present is known". For example let $T = 2$, and $\Lambda = \{3.14 < |X_3| < 3.15\}$, then (5.4) does NOT say that

$$P\{3.14 < |X_3| < 3.15 \mid 3 < |X_1| < 4; 3.1 < |X_2| < 3.2\}$$

$$= P\{3.14 < |X_3| < 3.15 \mid 3.1 < |X_2| < 3.2\};$$

namely that the past "$3 < |X_1| < 4$" may well have an after-effect on the future when the present $|X_2|$ is given as shown. Why? and what *does* (5.4) say? You must figure this out to understand what follows, correctly.[8]

We are going to use the Markov property to prove an important result on S_D, as follows: if D is a bounded Borel set then

$$\sup_{x \in R^3} E^x\{S_D\} < \infty \ . \tag{5.5}$$

For each $T = 1, 2, \ldots, n, \ldots$, the random variables X_n, $1 \le n \le T$, belong to $\mathcal{F}_T$, while $X_{T+1} - X_T$ belongs to $\mathcal{F}'_T$. Hence by an application of (5.4) and using the (abused) notation introduced in (3.9), we have

$$P\{X_n \in D \quad \text{for } 1 \le n \le T + 1\}$$

$$= P\{X_n \in D \quad \text{for } 1 \le n \le T; \quad P^{X_T}[X_1 \in D]\} \ . \tag{5.6}$$

Note that the temporal homogeneity of $\{X_t\}$ is used above as in (3.9). Note also that we have used P for E in the right member even though there is a random variable $P^{X_T}[\ldots]$ under it. This mild flip in notation will be indulged from time to time. Now by (3.7), we have

$$P^{X_T}[X_1 \in D] = \int_D p(1, X_T, y) dy \ . \tag{5.7}$$

For $X_T \in D$, this has the almost sure upper bound

$$\sup_{x \in D} \int_D p(1; x, y) dy = \theta < 1 \ . \tag{5.8}$$

To see why $\theta < 1$ it may be easiest to observe that when both x and y belong to D then $|x - y| \le M =$ the diameter of D, so that

$$\theta = \int_{|z| \le M} p(1; 0, z) dz < 1 \ .$$

Using this in (5.7), we obtain by induction on T in (5.6) that

$$\forall x \in D : \quad P^x\{X_n \in D \text{ for } 1 \le n \le T\} \le \theta^T \ . \tag{5.9}$$

Recall that the x in P^x indicates $X_0 = x$. Now the set $\{S_D > T\}$ is contained in the set $\bigcap\limits_{n=1}^{T} \{X_n \in D\}$ [warning: $S_D \geq T$ does not imply $X_T \in D!$]; hence

$$\forall x \in D : \quad P^x\{S_D > T\} \leq \theta^T . \tag{5.10}$$

For any random variable $S \geq 0$, its expectation satisfies the inequality

$$E(S) \leq \sum_{n=0}^{\infty} P(S > n) . \tag{5.11}$$

[This is true even if $P(S = 0) > 0$, why?] It follows that

$$\sup_{x \in D} E^x\{S_D\} \leq \sum_{n=0}^{\infty} \theta^n = \frac{1}{1 - \theta} < \infty . \tag{5.12}$$

How to extend the sup above to $x \in R^3$ is left to you as an exercise, it is easy but remember D is an arbitrary Borel set.[9]

An immediate consequence of (5.5) is that $S_D < \infty$ a.s. That is what we need for the immediate future. But let me pause to comment on the result in (5.5). For any Borel set D, we shall say it is "Green-bounded" when (5.5) holds; thus this is a condition on the set D that is broader (weaker) than ordinary Euclidean boundedness. Now S_D represents a random time (optional or not) and the sup of its expectation is a measure of the temporal size of D. Indeed, a little elaboration of the argument above yields an inequality valid in R^d, $d \geq 1$:

$$\sup_{x \in R^d} E^x\{S_D\} \leq A_d m(D) \tag{5.13}$$

where m denotes the Borel–Lebesgue measure and A_d is an absolute constant depending only on the dimension d (see [G]).

Many results valid for bounded sets can be extended to Green-bounded sets. As for the reason why it is called Green, it will be explained soon.

The next major step is to combine the temporal homogeneity hypothesis (3.7) with the Markov property in the form (5.3). For any $s \in R_0$, $t \in R_0$, we have for $f \in \mathcal{B}_b$, e.g.,

$$P_{s+t}f(x) = E^x\{f(X_{s+t})\} = E^x\{E^{X_s}[f(X_t)]\}$$
$$= E^x\{P_t f(X_s)\} = P_s(P_t f)(x) \qquad (5.14)$$

The second equation above holds as in (3.9) with $T = s$, and the rest is nothing but notation. The symbolism in $E^{X_s}[f(X_t)]$ is really horrid duplicity because if $s = t$ the two X_t in $E^{X_t}[f(X_t)]$ do not denote the same thing, as already noted under (3.9). No matter; the result may be recorded operationally as $P_{s+t} = P_s P_t$ and shows that $\{P_t, t \geq 0\}$ forms a "semigroup" with $P_0 =$ the identity operator. It is called the transition semigroup of the process. It is NOT the process, as some semigroupers would like to think.

We can express the joint distribution of any finite set of random variables from the process. It suffices to show this for four of them beginning with X_0. Thus let $0 = t_0 < t_1 < t_2 < t_3$. Then we have for $f_i \in \mathcal{B}_b$, $i = 1, 2, 3$:

$$E^x\{f_1(X_{t_1})f_2(X_{t_2})f_3(X_{t_3})\}$$
$$= P_{t_1}(f_1 P_{t_2-t_1}(f_2 \cdot P_{t_3-t_2}f_3))(x)$$
$$= \iiint p(t_1; x, x_1)f_1(x_1)p(t_2 - t_1; x_1, x_2)f_2(x_2)$$
$$\cdot p(t_3 - t_2; x_2, x_3)f_3(x_3)dx_1 dx_2 dx_3 \qquad (5.15)$$

where the integrals are all over R^3. Now the normal density function $p(t; x, y)$ given in (2.1) depends on (x, y) only through

$x - y$ (actually only $|x - y|$). Let us put

$$p_t(z) = p(t; , 0, z) = p(t; x, x + z) \qquad (5.16)$$

for any x; then the last member in (5.15) reduces to

$$\iiint p_{t_1 - t_0}(x_1 - x_0) p_{t_2 - t_1}(x_2 - x_1) p_{t_3 - t_2}(x_3 - x_2)$$

$$\times f_1(x_1) f_2(x_2) f_3(x_3) dx_1 dx_2 dx_3 \qquad (5.17)$$

where I have written x_0 for x. Noticing that $x_n = x_{n-1} + (x_n - x_{n-1})$, $n = 1, 2, 3$, this shows that the three random variables $X_{t_n} - X_{t_{n-1}}$, $n = 1, 2, 3$, are stochastically independent, and have respectively the probability densities $p_{t_n - t_{n-1}}$. Can there be any doubt that the same is true for any number of similar differences instead of three? Thus we have reached a fundamental property of the BMP, as follows.

> For any positive integer N and $0 = t_0 < t_1 < t_2 < \ldots < t_N$, the collection of $N + 1$ random variables X_0 and $X_{t_n} - X_{t_{n-1}}$ for $n = 1, 2, \ldots, N$, are stochastically independent with probability densities $p(t_n - t_{n-1}; 0, \cdot)$ given in Eqs. (5.16) and (2.1). $\qquad (5.18)$

In Einstein's investigation, he wrote [p. 13]:

> We will introduce a time-interval τ in our discussion, which is to be very small compared with the observed interval of time, but, nevertheless, of such a magnitude that the movements executed by a particle in two consecutive intervals of time τ are to be considered as mutually independent phenomena.

The editor R. Fürth appended a note to the above [p. 97]:

> The introduction of this time-interval τ forms a weak point in Einstein's argument, since it is not previously established

that such a time-interval can be assumed at all. For it might well be the case that, in the observed interval of time, there was a definite dependence of the motion of the particle on the initial state.

I have quoted the two passages above in order to show a sample of physicists' way of using mathematical language. If we take Einstein's words literally, the following conclusion will follow. Let $t_1 < t_2 < t_3$, and $t_2 - t_1 < \tau$, $t_3 - t_2 < \tau$, then clearly the two intervals (t_1, t_2) and (t_2, t_3) can be put into two contiguous intervals both of length $< \tau$, and so the movements in (t_1, t_2) and in (t_2, t_3) are mutually independent. It follows by induction that when any interval $(0, t)$ is divided into sufficiently small parts (of length less than τ), then the movements in all pairs of consecutive subintervals are independent. In Einstein's statement he considered only "two consecutive intervals" and said nothing about three or more such intervals. Nowadays any student of elementary probability knows [or does he?] that three events can be pairwise independent without being "totally" independent in the current usage of the word, see, e.g. [E; p. 142]. Apparently it was Serge Bernstein, teacher of Kolmogorov and Khintchine* *et al.* who first gave such a "counterexample". But Einstein probably did not care about such finesse and we must give him the benefit of doubt that by his mutual independence between consecutive intervals he really meant total independence among all the successive ones. If so, his hypothesis would be tantamount to (5.18) above. But then a *caveat* arises: there is no need of the τ to begin with.

---

*Apparently not in the literal sense, as A. Yu Veretennikov informed me, who also told of a case of inadequate domination forewarned on p. 9 above (March 2001).

The upshot seems to be: Einstein really had some more complicated model in mind which he did not make precise, of which the Brownian motion as formulated in (5.18) served as a first approximation.[10] Actually he wrote his first paper on the subject before he had heard about Brown, and his investigation "proved" the physical existence of molecules.

Let me also supply the hindsight regarding (2.9), a crucial step in Einstein's argument. This amounts to the fact that the (statistical) variance of $X(t)$ is equal to t, for all $t > 0$, or in the more vivid notation of stochastic calculus:

$$dX(t) = \sqrt{t} \, .$$

Observe that the t above is arbitrary, no τ there.

6. BROWNIAN CONSTRUCT

We are ready to apply (4.11) with $T = S_D$, where D is a Green-bounded domain. Since S_D is a finite optional time, Eq. (3.10) holds with $T = S_D$, namely:

$$Gf(x) = E^x \left\{ \int_0^{S_D} f(X_t)dt \right\} + P_{S_D}Gf(x) . \qquad (6.1)$$

This is the key to the Brownian resolution of Green's problem discussed in Sec. 1. The first term on the right-hand side above is an operator G_D defined as follows:

$$G_Df(x) = E^x \left\{ \int_0^{S_D} f(X_t)dt \right\} \qquad (6.2)$$

for suitable f, as always. It will be called the Green potential[11] operator for D. The expectation in (6.2) is defined for any domain D for a positive Borel function f, since the integral in (3.2) is then a well-defined $\mathcal{F}$-measurable function, though not necessarily finite and so by certain conventions not always qualified as a random variable. When $D = R^3$, then $S_D = +\infty$ and G_D reduces to G. Thus the "ground potential" is just an extreme case of the Green potential.

Next, let us examine the second term on the right-hand side of (6.1). By (4.12) we have

$$P_{S_D}(x, A) = P^x\{X(S_D) \in A\} . \qquad (6.3)$$

For each $x \in \bar{D}$, $P_{S_D}(x, \cdot)$ is a probability measure, ostensibly on the Borel tribe $\mathcal{B}^3$ (for R^3). But since the path $t \to X(t)$ is continuous by (4.5), at the first instant it leaves D it must be at the boundary, namely

$$\forall x \in \bar{D} : \qquad P^x\{X(S_D) \in \partial D\} = 1 . \qquad (6.4)$$

On the other hand, for each $x \notin \bar{D}$, we have $P^x\{S_D = 0\} = 1$ and so $P^x\{X(S_D) = X(0) = x\} = 1$. It seems dumb to bother with this banality, but it serves as a good reminder of the full force of the definitions. The last-written equation need not be true for $x \in \partial D = \bar{D} \backslash D$, owing to our choice of "$0 < t < \infty$" instead of "$0 \leq t < \infty$" in (5.1). This is a most serious matter which will turn up soon.

To repeat, when $x \in \bar{D}$ and $S_D < \infty$ as we are assuming, the probability measure defined in (6.3) is supported by the boundary ∂D. This measure is known outside probability theory as the "harmonic measure for D", arrived at usually through operatoric (functional-analytic) considerations. Let us re-denote it by $H_D(x, \cdot)$, restrict x to $\bar{D}$, and adopt a trendy notation as follows:

$$H_D(x, dz) = P^x\{X(S_D) \in dz\}, \qquad (x, z) \in \bar{D} \times (\partial D) . \quad (6.5)$$

Now take any bounded Borel function f defined on ∂D, written as $f \in \mathcal{B}_b (\partial D)$; then we have

$$H_D f(x) = E^x\{f(X(S_D))\}, \quad x \in \bar{D} . \qquad (6.6)$$

The function $H_D f$ is bounded because f is; is it measurable? For some psychological reason this kind of measurability question is regarded by many as a bore but it is inescapable, unfortunately. Indeed in probability theory we need a little more: the customary Lebesgue measurability is not serviceable, Borel measurability is necessary (why?). I will state a general result sufficient for our purposes, and indicate the steps of its proof, but leave the chores to you to carry out. After doing such exercises a few times, you may then wave hands next time around, may be.

First of all, we must sharpen our previous assertion that the function $(\omega, t) \to X(\omega, t)$ is $(\mathcal{F} \times \mathcal{B})$-measurable to its being $(\mathcal{F}_\infty \times \mathcal{B})$-measurable, where $\mathcal{F}_\infty$ is the tribe generated by all $\mathcal{F}_t, 0 < t < \infty$, namely by the process $\{X_t\}$. Then if $\omega \to T(\omega)$ is $\mathcal{F}_\infty$-measurable, $\omega \to X(\omega, T(\omega))$ is also $\mathcal{F}_\infty$-measurable. Any optional time T is $\mathcal{F}_\infty$-measurable, hence so is X_T. A set in $\mathcal{F}_\infty$ of the form, where N is a positive integer, $t_n \in R_0$, $A_n \in \mathcal{B}^3$:

$$\bigcap_{n=1}^{N} \{X_{t_n} \in A_n\} \tag{6.7}$$

is called a "cylinder set". For any such set Λ, the function $x \to P^x(\Lambda)$ is Borel measurable. For $N = 3$, this is seen by a direct inspection of (5.15), where we take $f_n = 1_{A_n}$; and the general case presents no problem. Now the cylinder sets generate the tribe $\mathcal{F}_\infty$, and Borel measurability is transmissible from them to the whole tribe. This is done by one of the "monotone class theorems" discussed in Sec. 2.1 of my Course [C]; the most convenient form to employ in this case, I think, is that given as Exercise 10 on p. 20 there, due to Sierpinski.[12] Thus the final result is as follows:

For any $\Lambda \in \mathcal{F}_\infty$, the function $x \to P^x(\Lambda)$
is Borel measurable (in R^3). (6.8)

As two consequences of (6.8): for any Borel measurable f, $G_D f$ defined in (6.2) and $H_D f$ defined in (6.6) are both Borel measurable.

Much ado about rather little! For immediate compensation we state the major fruition:

$$H_D f \text{ is harmonic in } D. \tag{6.9}$$

Fix an arbitrary x in D; the following statements are true P^x-a.s. Take any ball $B = B(x, r)$ such that $\bar{B} \subset D$. We know that the path starting at x will exit D for the first time at the time S_D, but before doing so it must exit from B, i.e., $S_B < S_D$, "stopping" momentarily at time S_B and place $X(S_B)$. By (4.10), the post-S_B process $\{X_t'\}$ behaves probabilistically exactly like the original process, but with the new starting place $X(S_B)$. *Its* path too will exit from D at *its* hitting time of D^c which is properly denoted by S_D'. We have by temporal homogeneity, for any y:

$$E^y\{f(S_D')\} = E^y\{f(S_D)\} = H_D f(y), \tag{6.10}$$

where in the first term the y in E^y indicates $X_0' = y$, whereas in the second term it indicates $X_0 = y$. (We could have used E'^y for the former.) Thus, by tracking the path from x via the "stop" at $X(S_B)$ until time S_D, we obtain

$$H_D f(x) = E^x\{f(X(S_D))\} = \int P^x\{X(S_B) \in dy\} E^y\{f(S_D')\}$$

$$= \int P^x\{X(S_B) \in dy\} H_D f(y)$$

$$= \int H_B(x, dy) H_D f(y) = H_B(H_D f)(x), \tag{6.11}$$

where in the last two equations we have used (6.5) with B for D. We know from our general discussion above that $H_B(x, \cdot)$ is supported by the 2-dimensional sphere $\partial B = \{y \in R^3 :$

$|y - x| = r\}$, but we can say a lot more. It is the uniform distribution on ∂B:

$$H_B(x, dy) = \frac{\sigma(dy)}{\sigma(\partial B)} \; . \qquad (6.12)$$

This is intuitively obvious from the spherical symmetry of the normal density (2.1), but a logical proof is more tedious than you might think because we are dealing with the random variable $\omega \rightarrow X(\omega, S_B(\omega))$. Using (6.12) and (6.11), we arrive at the desired relation

$$H_D f(x) = \frac{1}{\sigma(\partial B)} \int\limits_{\partial B} H_D f(y)\sigma(dy) \; , \qquad (6.13)$$

for any $x \in D$, and any ball B with center x and with closure contained in D. Such a function $H_D f$ is said to have the "sphere-averaging" property in D. If we now use polar coordinates and Fubini–Tonelli's theorem, we can integrate (6.13) along the radius to derive the "ball-averaging" property:

$$H_D f(x) = \frac{1}{m(B)} \int\limits_{B} H_D f(y) m(dy) \; . \qquad (6.14)$$

Note that we are using Lebesgue integration in two and three dimensions, respectively, in (6.13) and (6.14), and his big theorem that a bounded measurable function is always integrable.[13] If we did not know that $H_D f$ is (Borel) measurable, we could not even write down (6.13). Now it follows from (6.14) that $H_D f$ is a continuous function in D. This is a nice exercise, not hard, that you should try to do yourself; actually with a little trick one can even prove that any function having the sphere-averaging property, and integrable over *balls*, is infinitely differentiable in D, written as $C^\infty(D)$, but we do not need this here.

What have these properties to do with harmonicity? It was Gauss who established the sphere averaging property for a harmonic function; but it was later that Koebe (also Bôcher for R^2) showed the converse, more precisely: a continuous function which has the sphere-averaging property is harmonic (all in D).* The ancients did not know "measurable", but who needs it who had "continuous"? It is Koebe's part of this remarkable characterization of harmonicity that we need here to conclude that our Brownian construct $H_D f$ is harmonic in D for any $f \in \mathcal{B}_b(\partial D)$. As a matter of fact, one can even omit the boundedness of f, provided that the mathematical expectation in (6.6) be well-defined (what does this mean?) and that it be finite at some point (one is enough!) x in D. For this impressive generalization see [L; p. 154ff.].

---

*Koebe's theorem is an exercise in calculus using Gauss's divergence formula, a good companion to Green's formulas.

7. THE TROUBLE
WITH BOUNDARY

We now consider the boundary values of $H_D f$. Suppose we were to change the definition of S_D given in (5.1) one hair by replacing the first "$<$" in $\inf\{0 < t < \infty : \ldots\}$ with "$\leq$"; then for any $z \in \partial D$ we should have trivially

$$P^z\{S_D = 0\} = 1 \;, \tag{7.1}$$

and therefore $H_D f(z) = E^z\{f(X(0))\} = f(z)$. In other words, $H_D f$ would have the given boundary value f on ∂D. In particular, if f is continuous, i.e. $f \in C(\partial D)$, then since $H_D f \in C(D)$ and $H_D f = f$ on ∂D, it would be reasonable to expect that $H_D f \in C(\bar{D})$, namely that it is "continuous up to the boundary". Astonishingly, this is false in general.

The point of the "digression" (is it?) above is that even if we tried to cheat a little to "make life easier", the real problem would not go away. So we will return to the original definition of S_D and face the difficulty. Given D and a point z on its boundary ∂D, we shall say that D is "regular" at z, or that z is a regular boundary point of D,[14] in case

$$E^z\{S_D\} = 0 \tag{7.2}$$

which is equivalent to (7.1). We shall say that D is regular if D is regular at every $z \in \partial D$. A simple example to the contrary

43

is the punched ball $D = B(0, 1)\backslash\{0\}$, where 0 is an isolated boundary point. To see that D is not regular at 0, we need an important property of the BMP, valid in R^d, $d \geq 2$:

$$\forall x : \quad P^x\{T_{\{0\}} < \infty\} = 0 . \tag{7.3}$$

Recalling the definition in (4.2), this means the point $\{0\}$ is almost surely never hit. Owing to the spatial homogeneity of the process, the same is then true for any other point. We shall need this result again and it will be proved in Sec. 10. Accepting this for the moment, it is obvious that any path starting from 0 cannot hit $\partial D = \partial B(0, 1) \cup \{0\}$ right away, hence $S_D > 0$ a.s. This is much stronger than the negation of (7.2). It turns out that if D is not regular at $z \in \partial D$, then we must have

$$P^z\{S_D > 0\} = 1 . \tag{7.4}$$

This follows from a zero-or-one law due to R. M. Blumenthal, a doctorate student of G. A. Hunt's.

We are ready for the Boundary Convergence Theorem for a Green-bounded domain D, and $f \in \mathcal{B}_b(\partial D)$. If D is regular at $z \in \partial D$, and f is continuous at z, then we have

$$\lim_{x \to z} E^x\{|f(X(S_D)) - f(X(0))|\} = 0 \tag{7.5}$$

where x varies in $\bar{D}$. This is a form of the solution of Dirichlet's boundary value problem, apparently unnoted outside probability, which is actually equivalent to (7.16) below (why?).

The first step is to prove

$$\lim_{x \to z} E^x\{S_D\} = E^z\{S_D\} = 0 , \tag{7.6}$$

where now x may vary in all R^3 (why the difference between (7.5) and (7.6)?). Define an approximation of S_D, for each $\varepsilon > 0$:

$$S_D^\varepsilon = \inf\{\varepsilon < t < \infty : X_t \in D^c\} . \tag{7.7}$$

Then as $\varepsilon \downarrow 0$, $S_D^\varepsilon \downarrow S_D$ and

$$E^x\{S_D\} = \lim_{\varepsilon \downarrow 0} \downarrow E^x\{S_D^\varepsilon\} . \tag{7.8}$$

This requires a proof! Applying the post-ε temporal homogeneity, using (6.1) and observing that $E^x\{S_D\} = G_D 1(x)$ by (6.2), we obtain

$$E^x\{S_D^\varepsilon\} = E^x\{E^{X(\varepsilon)}[S_D]\} = P_\varepsilon(G_D 1)(x) . \tag{7.9}$$

Since $G_D 1$ is bounded by (5.5), this shows the function in (7.9) is also bounded. Its finiteness at x for some ε is sufficient for (7.8), by dominated (*sic*; not bounded or monotone) convergence.

Now is the occasion to announce another important property, this time not of the Brownian process, but *only* its transition probability, as follows.

For each $f \in \mathcal{B}_b(R^d)$ any $t > 0 :$ $P_t f \in C_b(R^d) .$ (7.10)

This is known as the "strong Feller property" — a misnomer because it is not stronger than another property called the "Feller property". The latter is fundamental for the construction of a class of Markov processes called "Feller process", precursor of Hunt process; see [L; Sec. 2.2]. The BMP is a special case of Feller process. It turns out that the conjunction of the two Feller properties, which I call "Doubly-Feller", possesses remarkable self-preserving qualities [Chung D]. Since (7.10) is a purely analytic result, its easy proof is left as an exercise.

Returning to (7.9), we see that for each $\varepsilon > 0$, $E^x\{S_D^\varepsilon\}$ is continuous in x. Hence for any z, we have

$$\overline{\lim_{x \to z}} E^x\{S_D\} \le \lim_{x \to z} E^x\{S_D^\varepsilon\} = E^z\{S_D^\varepsilon\} , \tag{7.11}$$

and then by (7.8) :

$$\overline{\lim_{x \to z}} E^x \{S_D\} \leq \lim_{\varepsilon \downarrow 0} E^z \{S_D^\varepsilon\} = E^z \{S_D\} \ . \qquad (7.12)$$

Since z is arbitrary, this means that as a function of x, $E^x \{S_D\}$ is upper semi-continuous in R^d. In general, it is not continuous. But under the condition (7.2), it is so at z, and that is what we need.

Heuristically, the result (7.6) should mean "in some sense" that S_D goes to 0. Suppose f is continuous on all ∂D, then since $X(\cdot)$ is continuous, it should follow that $f(X(S_D))$ goes to $f(X(0))$ "likewise". At least one book-author seemed to think so and left it at that.[15] "What's the problem?".

One problem has been forewarned in Sec. 4 in the definition of $X(S_D)$; it is $\omega \to X(\omega, S_D(\omega))$, a random function of a random function. It does not behave as nicely as say $\varphi(S_D(\omega))$ where φ is a non-random function. So the continuity of $t \to X(\omega, t)$ must be used with circumspection. Another more obvious problem is that we are dealing with the convergence of $E^x \{\dots\}$ as x varies, a kind of "weak convergence" but not quite. As far as I can see, there is no ready-made convergence-mode to enable us to draw (7.5), or (7.16) below, from (7.6). We have to go back to basics and put down a few epsilons and deltas.

Define a random (optional) time for each $r > 0$.

$$Z_r = \inf\{0 \leq t < \infty : \quad |X_t - X_0| > r\} \ . \qquad (7.13)$$

This is the first time that the path (or Brown–Einstein's particle) moves a distance r. By spatial homogeneity, we have for all $t \geq 0$:

$$P^x \{Z_r \leq t\} = P^0 \{Z_r \leq t\} \ . \qquad (7.14)$$

The exact distribution of Z_r is known but it is complicated and scarcely serviceable. Fortunately we need only the "mini info" that $Z_r > 0$ (a.s., why?). This is equivalent to

$$\lim_{t \downarrow 0} P^0\{Z_r \leq t\} = 0 \tag{7.15}$$

(why?). Now let $\varepsilon > 0$ be given; there exist three positive numbers r, τ, δ all depending on ε and each depends on the preceding one; such that

(i)
$$\sup_{\substack{|x-z|<r \\ |y-z|<r}} |f(y) - f(x)| < \varepsilon \ ;$$

(ii)
$$\sup_{|x-z|<r} P^0\{Z_r \leq \tau\} < \varepsilon \ ;$$

(iii)
$$\sup_{|x-z|<r} \sup_{|x-z|<\delta} P^x\{S_D > \tau\} < \varepsilon \ .$$

Continuity of f at z yields the r in (i); then (7.15) yields the τ; then (7.6) plus Chebyshev's inequality yields (iii).

Hence if $|x - z| < \delta$, then by (ii), (iii) and (7.14):

$$P^x\{S_D \geq Z_r\} \leq P^x\{S_D \geq Z_r; Z_r \leq \tau\} + P^x\{S_D \geq Z_r, Z_r > \tau\}$$

$$\leq P^x\{Z_r \leq \tau\} + P^x\{S_D > \tau\} < \varepsilon + \varepsilon = 2\varepsilon \ ;$$

consequently

$$E^x\{|f(X(S_D)) - f(X(0))|; \quad S_D \geq Z_r\} \leq 2M\varepsilon \ ,$$

where $M = \sup |f|$. If $|x - z| < r$, then we have by (i):

$$E^x\{|f(X(S_D)) - f(X(0))|; \quad S_D < Z_r\} < \varepsilon$$

because on the set $\{S_D < Z_r\}$, $|X(S_D) - X(0)| < r$.

Combining the two cases, we conclude that, if $|x-z| < \delta \wedge r$, then the sum of the two expectations is $< (2M + 1)\varepsilon$. This establishes the result (7.5). Is the argument too long? See Addenda I below.

A trivial Corollary is the usual form of the boundary convergence:

$$\lim_{x \to z} E^x\{f(X(S_D))\} = \lim_{x \to z} E^x\{f(X(0))\}$$

$$= \lim_{x \to z} f(x) = f(z) .$$

$$(7.16)$$

We have solved the classical Dirichlet boundary value problem. Let us denote it by (D, f), where D is a bounded and regular domain, and $f \in C(\partial D)$. The problem is to find a function h such that $h \in C(\bar{D})$, h be harmonic in D and $h \equiv f$ on ∂D. A solution is given by the $H_D f$ defined in (6.6). It is equal to f on ∂D. It is harmonic in D by (6.9); and as a special case of (7.16), it converges to f at every point of ∂D, hence it is continuous in $\bar{D}$.

In fact, $H_D f$ is the unique solution of (D, f). This is equivalent to the assertion that the identically zero function on $\bar{D}$ is the unique solution of the Dirichlet problem $(D, 0)$ where 0 is the identically zero function on ∂D. The proof follows from the ball-averaging property of a harmonic function given in Sec. 6, and is commonly referred to as the "maximum principle". Here is a large hint: how can the average of a ball of values be equal to its maximum or minimum? You should carry out the nice little argument yourself, purely analytic.

Dirichlet's problem may not be solvable without the condition of regularity (missed by Dirichlet as well as Gauss). The previous example of the punched unit ball will serve. Define $f = 1$ on $\partial B(0, 1)$ and $f(0) = 0$; then f is continuous! Both the functions h_1 and h_2, where $h_1(x) \equiv 1$, $h_2(x) \equiv \frac{1}{|x|}$, are harmonic in $D = B(0,1)\backslash\{0\}$, and have the correct boundary values on $\partial B(0,1)$, but miss being a solution at 0. A little trick (see Sec. 10 below) shows that there is no solution. Mathematicians at an earlier time must have disdained this example as not being a *regular* one (*pas normal*), but they had to accept a later construction by Lebesgue, known as a thorn or spine, as irrefutable. By the way, it was not so easy to *define* a regular boundary point as we did in (7.2), without Brownian theory, but it was done, by "barriers". The probabilistic solution was given first in 1944 by Shizuo Kakutani. He did it in R^2 and considered a Jordan domain (which is always regular); and instead of a continuous f he considered an arc on its boundary, so that he could construct the harmonic measure as given in (6.5), "in the sense of Nevanlinna". In his article he chose to omit some parts of the proof as being too delicate to be explained. As stated in Sec. 4, Wiener [D] proved the legitimacy of (4.5); a year later he [T] solved the generalized Dirichlet problem (which is essentially the assertion (7.16) for regular points, without regard to irregular points). But he did not employ the BMP in his work on the Dirichlet problem. There was hardly time and his notation did not provide room for a random time. He did give a necessary and sufficient condition for regularity, using Newtonian capacity which will be alluded to in Sec. 11. It is odd that in the later book by Paley and Wiener (1934), another proof of (4.5) was given by Fourier transform methods, but no mention was made of Wiener's 1923 paper [D].

8. RETURN TO GREEN

As reviewed at the beginning, Green was intent on the representation of a harmonic function by its boundary values, and used a heaven-made solution of a particular boundary value problem to obtain his formula. From the Brownian point of view the representation problem is easy! Let h be harmonic in a bounded and regular domain D. Surely Green would not mind it if we suppose that h is continuous up to the boundary, namely $h \in C(\bar{D})$. Then *pronto*!

$$\forall x \in D: \quad h(x) = \int_{\partial D} P^x \{X(S_D) \in dz\} h(z) \qquad (8.1)$$

is the representation sought. Why, it is just $h \equiv H_D h$ because the latter is the unique solution of the Dirichlet problem (D, h), whereas the given h is also a solution.

Observe that when D is a ball and x its center, (8.1) reduces to Gauss's spherical averaging. Thus Green's representation is a weighted averaging for a general D, given by the probability distribution of $X(S_D)$ — a big generalization of Gauss's theorem.

As a matter of fact, for a ball in any dimension, an explicit analytic formula is known as Poisson's. Let $B = B(0, r)$ in R^d, $d \geq 1$, and put

$$K_B(x, z) = \frac{1}{\sigma_d r} \frac{r^2 - |x|^2}{|x - z|^d}, \quad |x| < r, \quad |z| = r ; \qquad (8.2)$$

where

$$\sigma_d = \sigma(\partial B) = \frac{2\pi^{d/2}}{\Gamma\left(\frac{d}{2}\right)} .$$

Then we have

$$P^x\{X(S_B) \in dz\} = H_B(x, dz) = K_B(x, z)\sigma(dz) . \qquad (8.3)$$

When $d = 2$, and $x = (\rho, \theta)$ in polar coordinates, (8.1) becomes

$$h(\rho, \theta) = \frac{1}{2\pi} \int_0^{2\pi} \frac{r^2 - \rho^2}{r^2 - 2r\rho\cos(\theta - \varphi) + \rho^2} h(r, \varphi)d\varphi$$

$$= \frac{1}{2\pi} \int_0^{2\pi} \left\{ \frac{1}{2} + \sum_{n=1}^{\infty} \left(\frac{\rho}{r}\right)^n \cos n(\theta - \varphi) \right\} h(r, \varphi)d\varphi .$$

$$(8.4)$$

This can be derived from Cauchy's Integral Formula in complex function theory, which is a *representation* of a holomorphic f by its boundary values:

$$f(x) = \frac{1}{2\pi i} \int_C \frac{f(z)}{x - z}dz ,$$

see Titchmarsh [pp. 124–5]. I don't know if Green knew Cauchy's theorem. He did not mention it in his Essay.

When $d = 1$, $D = (-1, +1)$, ∂D is the two-point set $\{-1, +1\}$. It is remarkable that (8.2) still works and yields

$$K(x, -1) = \frac{1}{2}(1 - x), \quad K(x, +1) = \frac{1}{2}(1 + x), \qquad -1 < x < 1$$

$$(8.5)$$

with $\sigma(\{-1\}) = \sigma(\{+1\}) = 1$. To be explicit, we have

$$P^x\{X(S_B) = \mp 1\} = \frac{1}{2}(1 \mp x) \; ;$$

so that (8.1) becomes

$$h(x) = \frac{1}{2}(1-x)h(-1) + \frac{1}{2}(1+x)h(+1)$$

$$= \frac{h(+1) - h(-1)}{2}x + \frac{h(+1) + h(-1)}{2} \; ; \quad -1 < x < +1 \; .$$

This is the equation of the straight line passing through the two points $(-1, h(-1))$ and $(+1, h(+1))$. Euclid says there is exactly one such line. Since a harmonic function is a linear function in R^1, and vice versa, the calculations above amount to a probabilistic solution of the geometrical problem of representing the line, via Descartes' cartesian method. Quite a denouément. Since we did not prove (8.2) here, let me say that in R^1, the requisite Poisson kernel K_B, namely the evaluation in (8.5), can be done by probability argument (how?). Apropos, there is a challenging open problem regarding (8.2) for $d = 2$ and $d = 3$, or more specifically (8.4). As already said before, and clearly indicated in (8.3), the determination of $K_B(x, z)$ is that of calculating a nice distribution that has a simple geometric meaning. It ought to be solvable by direct methods of "geometrical probability" (formerly a well-known speciality, reported e.g. in the Encyclopedia Britannica), without the seemingly involuntary involvement with harmonicity *et cetera*. But apparently there is no easy way.

The proof indicated above for Green's representation (8.1) uses the full force of the unique solvability of the Dirichlet problem (D, h), hence it requires the regularity of D. Curiously, the latter hypothesis is dispensable so far as representation is concerned. Here is the argument well worthwhile for its general probabilistic flavor. Although D itself may not be regular, it can be approximated by regular subdomains. In fact standard geometrical consideration yields a sequence of "smooth" subdomains D_n such that $\bar{D}_n \subset D_{n+1}$ for all $n \geq 1$ and that $\bigcup_{n=1}^{\infty} D_n = D$, namely $D_n \uparrow\uparrow D$. The qualifier "smooth" here means that the surfaces ∂D_n have tangent planes *et cetera*. Such a domain can be shown to be regular. In fact, there is a more general criterion known as Poincaré–Zaremba's "cone condition", which has recently been flattened by a Brownian argument and thereby made considerably easier to apply, see [L; p. 165]. Anyway, we can apply (8.1) to each D_n to obtain a (the) representation of h in D_n, which we will write in probabilistic form:

$$\forall x \in D_n : \quad h(x) = E^x\{h(X(S_{D_n}))\} . \tag{8.6}$$

As $n \uparrow \infty$, $S_{D_n} \uparrow S_D$ (proof?), $X(S_{D_n}) \to X(S_D)$ and $h(X(S_{D_n})) \to h(X(S_D))$, by continuity of paths and continuity of h up to the boundary. Since h is bounded in $\bar{D}$, Lebesgue's bounded convergence theorem allows us to pass to the limit under E^x to obtain

$$\forall x \in D : \quad h(x) = E^x\{h(X(S_D))\} \tag{8.7}$$

which is (8.1). It seems odd that the regularity of D is not needed in this result. There is a hidden deep reason behind it. According to a general result of Kellogg–Evans–Doob–Hunt [L; p. 223], although there may be irregular points on ∂D, almost no path will ever hit them. Thus they are not really there so far as the paths are concerned. Earlier we mentioned that a singleton is never hit, which is a special case of this. A set so small that almost no path can hit it is called a "polar set"; it is also identifiable as a set of "Newtonian capacity" zero. Now that we have dropped this name, let us pause a moment to reflect on some consequences. Since each singleton is polar, and it is trivial that a countable union of polar sets is still polar, it follows that the set of all points in R^3 whose three coordinates are rational numbers is a polar set. They are everywhere dense, as you well know. Consequently almost no Brownian path will ever encounter any of them from time zero to eternity. Don't forget that all paths are totally continuous. How is it possible for them to twist and turn around the everywhere dense set without ever hitting it?

The above *is* a digression. Now let us return again to Green's own doing. Using the notation in Sec. 1, Green's special boundary value problem is the Dirichlet problem (D, g_y), where $y \in D$ and $g_y(x) = g(x, y) = \frac{1}{2\pi|x-y|}$ as in (2.3), and the solution is $\frac{1}{2\pi}h_y$ which is $H_D g_y$ in our new notation. The notation is confusing here owing to the duplicity of the functions involved which are all symmetric in (x, y). Let us write it out more explicitly:

$$\frac{1}{2\pi}h(x, y) = \int_{\partial D} H_D(x, dz)\frac{1}{2\pi|z - y|} . \qquad (8.8)$$

We now define the Green (or Green's) function for D as follows:

$$g_D(x, y) = \frac{1}{2\pi|x - y|} - \frac{1}{2\pi}h(x, y), \quad (x, y) \in D \times D . \qquad (8.9)$$

This was already spoken of in (1.10), but here the constant has been fixed. Be warned that other constants are adopted by other folks. You should check that the Green function is the density kernel of the Green operator in (6.2), namely:

$$G_D 1_A(x) = G_D(x, A) = \int_A g_D(x, y)dy \qquad (8.10)$$

for every Borel set A. For each x, the measure $G_D(x, \cdot)$ is supported by $\bar{D}$, as seen from (6.2).

Now let $y \in (\bar{D})^c$, then $u_y(\cdot) \in C(\bar{D})$ and u_y is harmonic in D. Therefore by the representation theorem just proved, we have $u_y = H_D u_y$ in D, namely

$$\frac{1}{|x-y|} = \int_{\partial D} H_D(x, dz) \frac{1}{|z-y|}, \quad x \in D, \quad y \in (\bar{D})^c . \quad (8.11)$$

This relation has an electric interpretation as follows. If a $+$ charge is placed at x inside the metal conductor surface ∂D, it induces $+$ charges on the outer side of ∂D and $-$ charges on the inner side of ∂D which are equal everywhere on ∂D, and their distributions are given by $H_D(x, dz)$. Since they cancel out nothing happens owing to their presence. When the conductor is *grounded*, however, all the $+$ charges on the outside of ∂D flow away to the ground, and only the $-$ charges on the inner side of ∂D remain. Now the original $+$ charge at x, combined with the induced $-$ charges distributed as $H(x, dz)$ on ∂D, produces a potential at each point y inside D, which is strictly positive and equal to $g_D(x, y)$ as shown in (8.9) — never mind the true physical constant instead of the $\frac{1}{2\pi}$ there. On the other hand, at each point y outside of D the two cancel out exactly and the phenomenon, easily observable experimentally, is mathematically shown in (8.11).

We are going to give a better derivation of the result (8.11) that is more intuitive and goes back to the fountain head. [What's intuition to one may be frustration to another!] Take a small ball $B(y, r)$ lying outside $\bar{D}$, and substitute its indicator 1_B for the f in (6.1). Reverting momentarily to the old notation P_{S_D} to remind us of the meaning of things, we have

$$\int\limits_B \frac{1}{|x-y|} dy = E^x \left\{ \int\limits_0^{S_D} 1_B(X_t) dt \right\}$$

$$+ \int\limits_{\partial D} P_{S_D}(x, dz) \int\limits_B \frac{1}{|z-y|} dy . \qquad (8.12)$$

For all $0 \le t < S_D$, we have $X_t \in D$, hence $X_t \notin B$ and $1_B(X_t) = 0$. Thus the middle term above is zero and (8.12) reduces to

$$\int\limits_B \frac{1}{|x-y|} dy = \int\limits_{\partial D} P_{S_D}(x, dz) \int\limits_B \frac{1}{|z-y|} dy . \qquad (8.13)$$

Now both members of (8.11), for each $x \in D$, are continuous functions of y in $(\bar{D})^c$. Therefore if we shrink B to its center y in (8.13), we obtain the desired (8.11).

In this new demonstration no "harmonic" was mentioned, and so it is independent of any results about harmonic functions. Only the post-S_D temporal homogeneity of the BMP and the continuity of its paths play the essential role. Hence the argument can be easily generalized to other processes and other potentials, such as Marcel Riesz's.

An explicit analytic formula for Green's function is rare. For a ball, however, a simple geometric trick due to Kelvin (Thomson), known as "inverse in a sphere", does it. For any $y \in R^3$, its inverse in the sphere $\partial B(0, r)$ is defined to be

$$y^* = \frac{r^2}{|y|^2} y . \qquad (8.14)$$

Thus y^* is the point on the line from 0 to y such that $|y||y^*| = r^2$. The Green function for $B = B(0, r)$ is then given by

$$g_B(x, y) = \frac{1}{2\pi} \left(\frac{1}{|x-y|} - \frac{r}{|y| \, |x-y^*|} \right) . \qquad (8.15)$$

To prove this it is sufficient, in view of (8.9), to show that the function $h(x, y)$ defined for each $y \in B$ by

$$h(x, y) = \frac{r}{2\pi|y|\,|x - y^*|} , \qquad x \in B \qquad (8.16)$$

is the solution of the Dirichlet problem $(B, g(\cdot, y))$. Since $y^* \in (\bar{B})^c$, $g(\cdot, y^*)$ is harmonic in B by (1.4), and $g(\cdot, y^*) \in C(\bar{B})$. Therefore by (8.16) the function $h(\cdot, y)$ has also these two properties. Now comes the geometry illustrated in the figure.

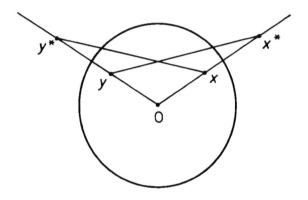

The two triangles $(0, x, y^*)$ and $(0, y, x^*)$ are similar, because $\overline{0x} \cdot \overline{0x^*} = \overline{0y} \cdot \overline{0y^*}$, hence

$$\frac{|x|}{|y|} = \frac{|x - y^*|}{|y - x^*|} = \frac{|y^*|}{|x^*|} . \qquad (8.17)$$

Consequently we have the remarkable symmetry below:

$$g_B(x, y) = g_B(y, x) . \qquad (8.18)$$

As a particular case of (8.17) when $x = z \in \partial B(0, r)$, so that $z = z^*$ and $|z| = r$, we have

$$\frac{|z|}{|y|} = \frac{|z - y^*|}{|y - z|} ,$$

which yields $h(z, y) = g(z, y)$. Thus $h(\cdot, y)$ has the assigned boundary values and so it is the aforesaid solution. This establishes the formula (8.15) — neat!

Following Green's foot-steps sketched in Sec. 1 but unscrambling the notation in (1.9), (1.10), (8.3) and (8.9), we arrive at Green's own representing kernel

$$-\frac{1}{2}\frac{\partial}{\partial n_y}g_B(x, y)\Big|_{y=z} \quad x \in B, \quad y \in B, \quad z \in \partial B. \quad (8.19)$$

Comparing this with (8.2) with $d = 3$, we conclude that the quantity above must be equal to

$$\frac{r^2 - |x|^2}{r|x - z|^3} \,. \quad (8.20)$$

To carry out the differentiation should be a soul-satisfying experience which is left to you. [I only checked the easy case for $x = 0$ to make sure that the negative sign in (8.19) in OK.]

My re-telling of the Green story must end here, but a few more words are in order for the amazing (8.18). Incredibly, this symmetry of Green's function persists for a general domain in any dimension. Did physicists such as the redoubtable Faraday discover this phenomenon experimentally, or did they possess the electrical perception to see that it must be so?[17] From our Brownian point of view, it is equivalent to the relation below:

$$E^x\left\{\frac{1}{|X(S_D) - y|}\right\} = E^y\left\{\frac{1}{|X(S_D) - x|}\right\}, \quad x \in D, \quad y \in D \,. \quad (8.21)$$

There does not seem to be any rhyme or reason for the paths to behave so reciprocally.

But we can derive the symmetry of Green's function from Green's identity (1.1)! Recall (8.9) but rewrite it as

$$G_y(x) = \frac{1}{2\pi|x-y|} - h_y(x) \qquad (8.22)$$

where $h_y(\cdot)$ is harmonic in D. Recall also the calculations leading from (1.5) to (1.6) which can be recorded symbolically, or in the distributional sense as

$$\Delta\frac{1}{|x|} = -4\pi\delta_0 \quad \text{or} \quad \Delta\frac{1}{|x-y|} = -4\pi\delta_y \ ; \qquad (8.23)$$

see Note. 2. It follows from (8.22) and (8.23) that

$$\Delta G_y = -2\delta_y - 0 = -2\delta_y \ . \qquad (8.24)$$

Since x and y are arbitrary points in D, we can of course interchange their roles in the above. Thus we have by (8.24)

$$\int_D G_x \Delta G_y dm = -2G_x(y) \ . \qquad (8.25)$$

Interchanging x and y we have

$$\int_D G_y \Delta G_x dm = -2G_y(x) \ . \qquad (8.26)$$

Now apply (1.1) with $u = G_x$, $v = G_y$; then both u and v vanish on the boundary because that is how Green constructed them. Therefore the right-hand member of (1.1) is zero, and the identity yields by (8.25) and (8.26):

$$-2G_x(y) + 2G_y(x) = \int_D (G_x\Delta G_y - G_y\Delta G_x)dm = 0 \ . \quad (8.27)$$

Thus we obtain $G_x(y) = G_y(x)$, the symmetry of $g_D(x,y)$ as claimed.

While the argument above lacks rigor in that it ignores the necessity of a smooth boundary and the existence of the normal derivatives in Green's identity, it should serve in electricity and magnetism to let us see the forest together with some of the leaves. Green could have done it. I wonder why he had to cite Professor Maxwell (see Note 17)?

9. STRONG MARKOV PROPERTY

The Markov property discussed in Sec. 5 is for a constant time T; it will now be extended to an optional time. For this purpose we need the pre-T tribe as well as the post-T tribe which has already been defined. How shall we describe an event which precedes the random time T? If T takes only rational values it is natural to consider an event of the form

$$\bigcup_{r \in \mathbb{Q}} \{T = r\} \cap \Lambda_r$$

where $\Lambda_r \in \mathcal{F}_r$. But if we wish to interpret the prefix "pre-" in the strict sense, then we should require $\Lambda_r \in \mathcal{F}_{r-} = \bigvee_{0 \le s < r} \mathcal{F}_s$.

Next, for a general T an approximation seems in order, but should we approximate it from the left or the right? It turns out in fact that there are a number of variations worth consideration. If you are curious you may consult [L, Sec. 1.3], or a fuller treatment in [O]. There are a lot of hair-splitting in the field!

We shall opt for a quick definition as follows. The pre-T tribe is the collection of subsets of $\mathcal{F}$ such that for each $t > 0$, we have

$$\Lambda \cap \{T < t\} \in \mathcal{F}_t . \tag{9.1}$$

We may even put $t = \infty$ in (9.1). Owing to the definition of optionality given in Sec. 4. the collection is a Borel tribe and is contained in $\mathcal{F}_\infty$. When T is a constant, it turns out to be $\mathcal{F}_{T+} = \bigwedge_{T < t < \infty} \mathcal{F}_t$. For this reason it will be denoted by $\mathcal{F}_{T+}$ in general. You should verify the preceding assertions; and also that both T and X_T are $\mathcal{F}_{T+}$-measurable.[18] The last result requires the (right) continuity of X_t in t.

Now we can state the extension of the Markov property in the form of (5.4).

For any finite optional time T, and $\Lambda \in \mathcal{F}'_T$:

$$P\{\Lambda \,|\, \mathcal{F}_{T+}\} = P\{\Lambda \,|\, X_T\} . \tag{9.2}$$

Combined with the post-option hypothesis in (4.11), this is called the Strong Markov Property.[19]

Up to here its full force has not been used. The main object of this section is to complete the picture and to give a simple but telling illustration.

We begin with the relation

$$e^{\int_0^{S_D} \varphi(X_t)dt} = e^{\int_0^{S_B} \varphi(X_t)dt} \, e^{\int_{S_B}^{S_D} \varphi(X_t)dt} \tag{9.3}$$

where S_D and S_B are the same times as in (6.11), and φ is a suitable function. It should be clear that the first exponential factor on the right-hand side of (9.3) belongs to the pre-S_B tribe, the second to the post-S_B tribe. Applying (9.2) in the "function" (instead of "set") form, we have

$$E\left\{ e^{\int_{S_B}^{S_D} \varphi(X_t)dt} \,\Big|\, \mathcal{F}_{S_B+} \right\} = E\left\{ e^{\int_{S_B}^{S_D} \varphi(X_t)dt} \,\Big|\, X(S_B) \right\} \tag{9.4}$$

Applying the post-S_B homogeneity, we see that the right-hand member of (9.4) is equal to

$$E^{X(S_B)} \left\{ e^{\int_0^{S_D} \varphi(X_t)dt} \right\}. \tag{9.5}$$

If you do not understand why the $\int_{S_B}^{S_D}$ there has become $\int_0^{S_D}$ here, you should re-read the explanations leading to (6.11). The situation has become a little more complicated but the idea is the same. Let us now take the indicated conditional expectations of both members of (9.3):

$$E^x \left\{ e^{\int_0^{S_D} \varphi(X_t)dt} \,\Big|\, \mathcal{F}_{S_B+} \right\}$$

$$= e^{\int_0^{S_B} \varphi(X_t)dt} E^x \left\{ e^{\int_{S_B}^{S_D} \varphi(X_t)dt} \,\Big|\, \mathcal{F}_{S_B+} \right\}$$

$$= e^{\int_0^{S_B} \varphi(X_t)dt} E^{X(S_B)} \left\{ e^{\int_0^{S_D} \varphi(X_t)dt} \right\}, \tag{9.6}$$

by (9.4) and (9.5). Put (with apology to the previously used u)

$$u(x) = E^x \left\{ e^{\int_0^{S_D} \varphi(X_t)dt} \right\}. \tag{9.7}$$

Taking unconditional expectation in (9.6), we obtain

$$u(x) = E^x \left\{ e^{\int_0^{S_B} \varphi(X_t)dt} \, u(X(S_B)) \right\} . \tag{9.8}$$

When φ is a negative number $-\lambda$, $\lambda > 0$, we have

$$u(x) = E^x \left\{ e^{-\lambda S_D} \right\} , \tag{9.9}$$

the Laplace transform of S_D which uniquely determines the probability distribution of S_D.

What does (9.8) signify? To appreciate it we need go a step farther as follow: take a suitable f and insert $f(X(S_D))$ as a posterior factor on both sides of the relation (9.3), and then go through the operations above, which requires no new argument. Now put

$$u_f(x) = E^x \left\{ e^{\int_0^{S_D} \varphi(X_t)dt} \, f(X(S_D)) \right\} . \tag{9.10}$$

Then we obtain exactly the same result (9.8) with u_f replacing the u there. (Isn't it odd that the posterior factor has no effect on the equation?) With this amplification, we obtain

$$u_f(x) = E^x \left\{ e^{\int_0^{S_B} \varphi(X_t)dt} \, u_f(X(S_B)) \right\} . \tag{9.11}$$

When $\varphi \equiv 0$, this reduces to an old acquaintance. Find it.

Instead of a ball B in the above, we may take any subdomain with its closure contained in D. Then if $u_f \in C(\bar{D})$, (9.11) will hold even with the B there replaced by D, by the same argument in the passage from (8.5) to (8.6). Thus we get a representation of u_f by its boundary values in a certain sense. Observe that in the exponential factor the past of the process is carried up to the time of exit from the boundary. It is by no means evident that as x goes to the boundary that past will leave no "after effect" on the limit.

We have delayed a vital question here: What is a suitable φ in (9.7)? To see the seriousness of this question consider $\varphi \equiv 1$: then $u(x) = E^x\{e^{S_D}\}$. Of course by Fubini–Tonelli we have

$$E^x\{e^{S_D}\} = \sum_{n=0}^{\infty} \frac{1}{n!} E^x\{S_D^n\}$$

and if D in Green-bounded then $E^x\{S_D\}$ is bounded (by definition); but what about those higher powers? Thus even if we suppose D and φ to be both bounded, we are far from knowing if the u in (9.7) is finite. To stress this point, suppose we replace the random time S_D by a constant time T. Then if $|\varphi|$ is bounded by M we have trivially

$$E^x\left\{ e^{\int_0^T \varphi(X_t)dt} \right\} \leq e^{MT} \ .$$

But can we substitute S_D for T in the above by making use of the distributional properties of S_D? We know for instance that $P^x\{S_D > T\}$ decreases exponentially fast as $T \uparrow \infty$ (the famous "exponential decay" of radio-activity!), and after all, the exponential function in the formula is "simple stuff". Years ago I asked Rick Durrett (who was then a student in my class) to investigate the possibilities. He made a valiant attempt to concoct a large-deviation connection. It did not work. I was informed more recently that a third stage of the deviationary tactics was still in the offing. Anyway it is clear that there is a world of difference between a constant T and the random S_D, exactly the difference between a semigroup and the stochastic process, already spoken of in Sec. 4. In fact, except in the banal case where $\varphi \leq 0$, which was much studied by Mark Kac, we know of no viable condition to ensure that $u(x) < \infty$. On the other hand, there are necessary and sufficient conditions on D and φ for this. Sounds odd? Not really because "necessary and sufficient" means tautological and therefore may just give us a run-around. Instead, I will now give the following viable answer.

Let D be a bounded domain (in R^d, $d \geq 1$) and let φ satisfy the condition below due to Aizenman and Simon:

$$\lim_{t \downarrow 0} \sup_{x \in R^d} \int_0^t P_s|\varphi|(x)ds = 0 . \tag{9.12}$$

Then we have the following all-or-nothing criterion. Either the function u in (9.7) is equal to $+\infty$ everywhere in D, or it is bounded not only in D but in $\bar{D}$. This function is called the *gauge* for (D, φ).

Consequence: In case the gauge is bounded, it is continuous and > 0 in D, and satisfies the equation

$$\left(\frac{\Delta}{2} + \varphi\right) u = 0 \tag{9.13}$$

in the "weak" or distributional sense. Furthermore, the function u_f in (9.11) also satisfies the equation (9.13); and if D is regular then $u_f \in C(\bar{D})$ and it is the unique function having these properties, namely u_f is the unique solution of the Dirichlet boundary value problem for the new equation (9.13). Note that the gauge is just u_1. Note also that when φ is the zero function we are back to our previous occupation.

It is trivial that any bounded Borel function φ satisfies the condition (9.12). It is more fun to check that the famous Newton–Coulomb function $|x|^{-1}$ also satisfies the condition.

The equation (9.13) is called Schrödinger's equation but in truth it is only a *real* fake because he had a factor $i = \sqrt{-1}$ in front of the φ in (9.13), as well as some physical constant (Planck's?) which makes it "purely imaginary" — called "quantum". Probability theory still lives in the world of reals and does not know how to handle that imaginary stuff. In 1949 I heard a talk by Richard Feynman[20] on this topic, and the exponential $e^{\int_0^t \varphi(X_s)ds}$ is now commonly known as Feynman's "path integral", which is a double misnomer because firstly he (like Schrödinger) had an $i = \sqrt{-1}$ there; secondly, what he called path integral is nothing but our elementary $E = P$ over the path space $\{\omega \in \Omega : \ X(\omega, \cdot)\}$, not the true path-by-path integral $\int_0^t \varphi(X_s(\omega))ds$. By the way, he considered only constant t there but quite possibly would not mind randomizing it when his calculations warranted it.

The story is too long to tell here. Suffice it to say that for a "gaugeable" pair (D, φ) a probabilistic theory has been developed for the Schrödinger extension of the Laplacian *ideenkreis* begun in Sec. 1 of this book. A short exposition has been given in Sec. 4.7 of [L]; a fuller one in [F] which has just been published.

With the enunciation of the strong Markov property we have completed the gradual unveiling of the character of the Brownian Motion process. It is a strong Markov Process with the transition density function given by (2.1) in the Euclidean space R^d, and with continuous sample paths. The "strong" assertion actually follows from the rest of the characterization.

The older characterization embodied in (5.18) can be shown to be an equivalent one owing to the spatial homogeneity of the transition density; see Sec. 4.1, Theorem 4 of [L].

10. TRANSIENCE

The BMP in R^3 has a property radically different from that in R^1 or R^2. It has been given the name "transience" or "non-recurrence" but it can be more simply described by:

$$P\left\{\lim_{t\to\infty} |X(t)| = +\infty\right\} = 1 \ . \qquad (10.1)$$

This result has a famous discrete analogue originating in Pólya's Bernoullian random walk (1921), and extended to a general random walk, see [C; Sec. 8.3]. The latter implies at once the following property of the BMP in R^d, $d \geq 1$. For each $r > 0$ and $\delta > 0$, we have

$$P\{|X(n\delta)| > r \quad \text{for all sufficiently large } n\}$$
$$= \begin{cases} 0, & \text{if } d = 1 \text{ or } 2; \\ 1, & \text{if } d \geq 3. \end{cases}$$

For $d = 1$ or 2, this is a stronger assertion than

$$P\{|X(t)| > r \quad \text{for all sufficiently large } t\} = 0 \ ;$$

while for $d \geq 3$, it is a weaker result than

$$P\{|X(t)| > r \quad \text{for all sufficiently large } t\} = 1 \ .$$

Thus (10.1) does not follow from the discrete analogue

$$\forall \delta > 0 : \quad P\{\lim_{n\to\infty} |X(n\delta)| + \infty\} = 1 \ , \qquad (10.2)$$

which can of course be immediately strengthened to a countable set of δ such as the rational numbers. One would expect to be able to squeeze out from the latter a proof of (10.1) by approximation and the continuity of $X(\cdot)$. Surprisingly, there is no easy way. First of all, it is "easy" to construct a continuous f from R_0 to R^1 such that $\lim_{n\to\infty} f(nq) = +\infty$ for every rational q, while $\lim_{t\to\infty} f(t) = 0.$[21] Therefore some special property of the BMP is indispensable to get from (10.2) to (10.1). But how much do we need? If the strong Markov property is used, this can be done by *reductio ad absurdum*, using (7.15) with $r = 1$, and a form of the Borel–Cantelli lemma similar to Theorem 9.5.2 of [C], which Doeblin called the "street crossing lemma". This argument requires only $\lim_{n\to\infty} |X(n)| = +\infty$, *viz.* $\delta = 1$ in (10.2).

In general, it would be foolish to treat the BMP by approximating it with discrete skeleton — a remark made already by Bachelier who initiated the continuous time model (for stock market "speculation"). Let us therefore go to prove (10.1) by the new methodology. We begin by calculating the probability of hitting one set before another. Let $0 < a < b < \infty$; $B_a = B(0, a)$, $B_b = B(0, b)$, $D = B_b \backslash \bar{B}_a$. Then D is a domain with $\partial D = (\partial B_a) \cup (\partial B_b)$; it is regular because spheres are as smooth as they come and satisfy the cone condition, although we have not proved this. Define f to be equal to one on ∂B_a and equal to zero on ∂B_b; this is a continuous function on ∂D. Hence we know from Sec. 7 that the following function h is harmonic in D:

$$h(x) = E^x\{f(X(T_{D^c}))\} = P^x\{T_{\partial B_a} < T_{\partial B_b}\} \qquad (10.3)$$

where T_A is defined in (4.2). In fact, h is the unique solution of the Dirichlet problem (D, f). It is geometrically obvious that

h depends only on $|x|$ (not on the angle). Now the Laplace equation (1.2) for a function of $r = \sqrt{x_1^2 + x_2^2 + x_3^2}$ takes the form

$$\frac{d^2h}{dr^2} + \frac{2}{r}\frac{dh}{dr} = 0 \ . \tag{10.4}$$

This ordinary linear differential equation is easy to solve and has the general solution $\frac{c_1}{r} + c_2$ where c_1 and c_2 are constants. By (10.3), and the regularity of D, we have $h(r) = 1$ or 0 for $r = a$ or b. It follows that

$$h(x) = \frac{ab}{(b-a)|x|} - \frac{a}{b-a}, \qquad a \le |x| \le b \ . \tag{10.5}$$

Letting $b \uparrow \infty$, so that $T_{\partial B_b} \uparrow \infty$, we obtain

$$P^x\{T_{\partial B_a} < \infty\} = \frac{a}{|x|}, \qquad a \le |x| < \infty \ . \tag{10.6}$$

This formula can be extended to all x if we truncate the right-hand quantity by 1; and $T_{\partial B_a}$ can be replaced by $T_{\bar{B}_a}$ or T_{B_a}.

Now for a general set A, put

$$\varphi_A(x) = P^x\{T_A < \infty\} \ . \tag{10.7}$$

We have then

$$P_t\varphi_A = E^x\{\varphi_A(X_t)\} = E^x\{P^{X_t}[T_A < \infty]\}$$
$$= P^x\{\exists s \in (t, \infty) : X_s \in A\} \ ; \tag{10.8}$$

and consequently by (4.3):

$$\lim_{t\to\infty} P_t\varphi_A(x) = P^x\{L_A = +\infty\} \ . \tag{10.9}$$

Whether it is by nature's grand design or by pure chance, the mathematics required here will lead us straight back to the original source in Sec. 1 and Sec. 2. Recall first the Newton–Coulomb potential $u_0(x) = |x|^{-1}$ in (1.3), so that the right member of (10.6) is au_0; while the left member has been re-denoted by $\varphi_{\bar{B}_a}(x)$. Next recall P_t from (2.5), and the quintessential representation of u_0 in (2.3). We have then

$$P_t u_0(x) = \int_{R^3} p(t; x, y) \left[\int_0^\infty p(s; y, 0) ds \right] dy$$

$$= \int_0^\infty \left[\int_{R^3} p(t; x, y) p(s; y, 0) dy \right] ds$$

$$= \int_0^\infty p(t + s; x, 0) ds = \int_t^\infty p(s; x, 0) ds .$$
$$\tag{10.10}$$

Of course, it is Fubini–Tonelli that is used first, then the convolution property of the normal density, which makes $\{P_t\}$ a semigroup — every step is fundamental! Letting $t \to \infty$ we obtain by (10.9) with $A = \bar{B}_a$:

$$P^x \{ L_{\bar{B}_a} = \infty \} = 0, \qquad x \in R^3 . \tag{10.11}$$

Since any bounded set A in R^3 is a subset of B_a for some a, we may replace $\bar{B}_a$ in (10.11) by A. But if A is not Borel measurable, not even an analytic set, L_A may not be a random variable, how can we have $P^x \{ L_A = \infty \}$? Well, our probability space is complete (see [C; p. 383ff.]); if $A \subset B_a$ then $L_A \leq L_{B_a}$, $\{ L_A = \infty \} \subseteq \{ L_{\bar{B}_a} = \infty \}$. By completeness, when (10.11) is true, then $P^x \{ L_A = \infty \}$ is defined and is also equal to 0. Therefore, it is logical to speak of the transience of an arbitrary bounded set A.

There is another big consequence of (10.6). For any $x \neq 0$, if $a \downarrow 0$ then $T_{\partial B_a} \uparrow T_{\{0\}}$, hence

$$P^x \{ T_{\{0\}} < \infty \} = 0 . \tag{10.12}$$

This result remains true for $x = 0$, an excellent little exercise. It has been announced before in (7.3): each singleton is a polar set. This result is also true in R^2, but totally false in R^1. The former assertion can be proved in a similar way as in R^3 shown above. The latter is "easy to see", as Laplace used to say in his book dedicated to Napoleon.

11. LAST BUT NOT LEAST

Let us now fix a compact set A and consider L_A. So far we have ignored its measurability which is tied to that of T_A, as follows. For each $s > 0$, define

$$T_A^s = \inf\{s < t < \infty : \quad X_t \in A\} . \tag{11.1}$$

A similar random time has already been used in (7.7). This T_A^s is the T_A for the post-s process $\{X_t^s\} = \{X_{s+t}\}$ introduced in Sec. 3. We have seen in Sec. 4 that for a closed A, T_A is optional and so $\mathcal{F}_\infty$-measurable. Hence T_A^s is $\mathcal{F}_\infty^s$-measurable, where $\mathcal{F}_\infty^s = \sigma(X_t, s \leq t < \infty)$. Since $\mathcal{F}_\infty^s \subset \mathcal{F}_\infty$, T_A^s is a random variable (allowing $+\infty$ as value). Now the definition of L_A implies that for each $s > 0$, we have

$$\{L_A > s\} = \{T_A^s < \infty\} . \tag{11.2}$$

It follows firstly that L_A is $\mathcal{F}_\infty$-measurable; secondly that

$$\{L_A > 0\} = \{T_A < \infty\} ; \tag{11.3}$$

thirdly that

$$\{0 < L_A \leq s\} = \{T_A < \infty\} - \{T_A^s < \infty\} ; \tag{11.4}$$

and fourthly for each x:

$$P^x\{0 < L_A \leq s\} = P^x\{T_A < \infty\} - P^x\{P^{X_s}[T_A < \infty]\}$$
$$= \varphi_A(x) - P_s\varphi_A(x) \tag{11.5}$$

where φ_A is already introduced in (10.7).

The function φ_A is harmonic in A^c. To see this let $x \in A^c$ and $\overline{B(x,r)} \in A^c$. Then using an old argument in Sec. 6, and writing ∂B for $\partial B(x,r)$:

$$\varphi_A(x) = P^x\{T_A < \infty\} = P^x\{T_{\partial B} < T_A < \infty\}$$
$$= P^x\{P^{X(T_{\partial B})}[T_A < \infty]\} = H_B\varphi_A(x) \ .$$
$$(11.6)$$

Namely φ has Gauss's sphere-averaging property, and being bounded, is harmonic in A^c. What can we say about $\varphi_A(x)$ for $x \in A$? If $A = \bar{D}$ where D is a bounded domain then it is trivial that $\varphi_A(x) = 1$ for $x \in D$; if $(\bar{D})^c$ is regular as defined in (7.2), namely if $E^x\{T_{\bar{D}}\} = 0$ for $x \in \partial D$, the same is true for $x \in \bar{D}$.

It follows that each $x \in A^c$, there is a neighborhood of x in which φ_A belongs to $C_b^{(2)}$, hence by a theorem on p. 190 of [L],

$$\lim_{s \to 0} \frac{\varphi_A(x) - P_s\varphi_A(x)}{s} = -\frac{\Delta}{2}\varphi_A(x) = 0 \ . \qquad (11.7)$$

In the case just mentioned, where φ_A is the constant 1 in $A^0 = D$ (A^0 is the interior of A), the same conclusion holds if $x \in A^0$. Thus when $A = \bar{D}$, the limit in (11.7) is 0 except possibly for $x \in \partial D$.

Previous Secs. 5 to 9 are all about T_{D^c} and $X(T_{D^c})$. "Cherchez la symétrie"! So let us now find out about $L_{\bar{D}}$ and $X(L_{\bar{D}})$. As before we write $A = \bar{D}$ for a bounded domain D, to "fix the ideas" as they say, but remark that in much of the sequel A may be an arbitrary compact set in R^3. We proved in Sec. 10 that $L_A < \infty$ (a.s.). Is there any chance to say more? It follows from (11.2) that

$$E^x\{L_A\} = \int_0^\infty P^x\{L_A > t\}dt = \int_0^\infty P_t\varphi_A(x)dt$$

$$= G\varphi_A(x) \ .$$

If $A = \overline{B(0,a)}$, then by (10.6), we have $\varphi_A(x) = \frac{a}{|x|} \wedge 1$ and so the above is equal to

$$\int_{R^3} \frac{1}{2\pi|x-y|} \left(\frac{a}{|y|} \wedge 1 \right) = +\infty \ .$$

Thus $E^x\{L_A\} = \infty$. Contrast this with (5.12): $\sup\limits_{x} E^x\{T_{A^c}\} < \infty$. Indeed the latter can be further strengthened to $\sup\limits_{x} E^x\{\exp(bT_{A^c})\} < \infty$ for some $b > 0$.

Let us turn to $X(L_A)$, and put

$$L_A f(x) = E^x\{L_A > 0; f(X(L_A))\} \ . \tag{11.8}$$

Just as $X(T_{D^c}) \in \partial D$ by continuity of the Brownian paths, so we have $X(L_A) \in \partial A = \partial D$ when $A = \bar{D}$. Thus it is sufficient in (11.8) to consider $f \in C(\partial A)$ to determine the distribution of the last exit place. However, we shall need to use $f \in C_c(R^3)$ below. [There is an old theorem by Lebesgue asserting that any function given on a compact set can be extended to a continuous function on the whole space.] Anyway, to "get hold of" the random variable $X(L_A)$ we will "function it" and then approximate from the left[22] as follows, writing L for L_A:

$$f(X_L) = \lim_{\delta \downarrow 0} \frac{1}{\delta} \int_{(L-\delta)^+}^{L} f(X_s)ds \ , \tag{11.9}$$

where $(L-\delta)^+ = (L-\delta) \vee 0$. Now re-write the integral above as

$$\int\limits_{0}^{\infty} f(X_s)1_{\{s<L\leq s+\delta\}}ds \qquad (11.10)$$

so that we are in position to play the Fubini–Tonelli game. If we now take expectation and change the order of integration, we obtain

$$\int\limits_{0}^{\infty} E^x\{f(X_s)1_{\{s<L\leq s+\delta\}}ds\} . \qquad (11.11)$$

See how the vital but tedious proviso "$L > 0$" in (11.8) has bowed out automatically? We are ready to apply the Markov property for further revelation. Using (11.6) we obtain from (11.11):

$$\int\limits_{0}^{\infty} E^x\left\{f(X_s)E^{X_s}(0 < L \leq \delta)\right\}ds$$

$$= \int\limits_{0}^{\infty} E^x\left\{f(X_s)(\varphi_A - P_\delta\varphi_A)(X_s)\right\}ds$$

$$= E^x\left\{\int\limits_{0}^{\infty} f(\varphi_A - P_\delta\varphi_A)(X_s)ds\right\}$$

$$= G(f(\varphi_A - P_\delta\varphi_A))(x) , \qquad (11.12)$$

a small miracle of symbolism bringing us back to (3.3). Since f has bounded support and $0 \leq \varphi_A \leq 1$, we have

$$G(|f| |\varphi_A - P_\delta\varphi_A|) < \infty ,$$

in fact bounded by an old exercise in Sec. 2. Therefore, the calculations above are correct by Fubini–Tonelli, and upon

retracing the steps we see that we have proved the bewildering
result below:

$$L_A f = \lim_{\delta \downarrow 0} G\left(f \cdot \frac{\varphi_A - P_\delta \varphi_A}{\delta}\right) . \qquad (11.13)$$

Actually we have also interchanged the order of the indicated
limit in δ with E^x, which must be justified by bounded convergence, namely:

$$\left| \frac{1}{\delta} \int\limits_{(L-\delta)^+}^{L} f(X_s) ds \right| \le \|f\| .$$

Put for all Borel sets B in R^3:

$$\mu_\delta(B) = \int\limits_{B} \frac{1}{\delta}[\varphi_A(y) - P_\delta \varphi_A(y)] dy . \qquad (11.14)$$

Then we can re-state (11.13) as follows, for $f \in C_c(R^3)$ and
$x \in R^3$:

$$\lim_{\delta \downarrow 0} \int\limits_{R^3} \frac{f(y)}{2\pi|x - y|} \mu_\delta(dy) = L_A f(x) . \qquad (11.15)$$

From this key relation we can deduce, without difficulty, the
conclusions below.

(a) As $\delta \downarrow 0$, the measure μ_δ converges vaguely to a measure
μ_A defined as follows: for any $x \in R^3$:

$$\mu_A(dy) = 2\pi|x - y|L_A(x, dy) \qquad (11.16)$$

where the measure $L_A(x, dy)$ is that indicated (*à la Bour-
baki*) by (11.8), with support in ∂A. That μ_A has support
∂A has also been confirmed by the remark following (11.7).

(b) The measure μ_A is atomless, namely $\mu_A(\{y\}) = 0$ for each $y \in R^3$.

(c) Turning the table around, we have for each $x \in R^3$:

$$L_A(x, dy) = \frac{\mu_A(dy)}{2\pi|x - y|}, \quad y \neq x,$$

(11.17)

$$L_A(x, \{x\}) = 0.$$

The ultimate equation above is worthy of comment. While $L_A(x, \{y\}) = 0$ for $y \neq x$ follows from (b), that single excluded case seems to require (10.2) in the form $P^x\{T_{\{x\}} < \infty\} = 0$, which was left dangling there for a tease.

12. LEAST ENERGY

Summing up the results in the preceding section, we have for each compact subset A in R^3, a measure μ_A supported by ∂A with the following property. For each $f \in C(\partial A)$ and $x \in R^3$:

$$E^x\{L_A > 0; \; f(X(L_A))\} = \int_{\partial A} g(x,y)f(y)\mu_A(dy) \qquad (12.1)$$

where g is given in (2.3). Introduce the notation

$$G\nu(x) = \int g(x,y)\nu(dy) \qquad (12.2)$$

for any measure ν, as a natural extension of Gf for a function f. Putting $f \equiv 1$ in (12.1) and recalling (11.4), we obtain the celebrated relation

$$P^x\{T_A < \infty\} = G\mu_A(x) . \qquad (12.3)$$

The probability on the left side above is of course equal to 1 for $x \in A^0$; it is also equal to 1 for x in $\partial A = \partial(A^c)$ which is regular as a boundary point of the open set A^c, defined in (7.1). We will assume this below for all $x \in \partial A$ so that (12.3) includes

$$\forall x \in A : \quad G\mu_A(x) = 1 . \qquad (12.4)$$

The measure μ_A is known as the "equilibrium measure" for A, and has the electrostatic origin as nature's manner of distributing a charge placed on the conductor A. The total charge is of course $\mu_A(A)$ which will be re-denoted by $\mu_A(1)$, conformable with the general notation $\mu_A(f) = \int f d\mu_A$ for a function f. The potential function produced by this charge, under the Newton–Coulomb law, is clearly given by $G\mu_A$, as a weighted average of $g(\cdot, y)$ with y distributed according to μ_A.

This is constant throughout the conductor A, where the maximum potential is achieved, as seen from (12.3). Electrostatically speaking, the constant 1 in (12.4) amounts to a unit-taking. In this sense the total charge $\mu_A(1)$ is called the "capacity" of the conductor A and denoted by $C(A)$. This is the maximum charge that the conductor A can support so that the resultant potential everywhere in space not exceed one unit. Hence the name "capacity", but I do not know how Sir Isaac got it before electricity was invented.

It is possible to extend this notion of capacity to all "capacitable" sets. As mentioned earlier in Sec. 4, Hunt used such capacities to prove that for all analytic, hence all Borel sets A, the random variable T_A is an optional time. For the present mini discussion let us just note that $C(A) = C(\partial A)$ although A may be 3-dimensional and ∂A 2-dimensional. It is easy to deduce from (12.3) that for the solid ball $A = \overline{B(0, r)}$ or its surface, we have $C(A) = 2\pi r$. This can also be deduced from the less informative relation (12.4) if we take $x = 0$ there.[23] In fact, the rotational symmetry of the BMP implies that the distribution $L_A(0, \cdot)$ should be uniform over ∂A; hence by (11.16):

$$\mu_A(dy) = 2\pi r \frac{\sigma(dy)}{\sigma(\partial A)} = \frac{1}{2}\sigma(dy) \ .$$

$$\mu_A(\partial A) = \frac{1}{2}\sigma(\partial A) = 2\pi r \ .$$

The existence of the equilibrium measure is a physical phenomenon, just as the induced charge distribution on Green's conductor recounted in Sec. 1. As shown in (6.3) and (6.5), the latter, the harmonic measure $H_D(x, \cdot)$, is nothing but the probability distribution of $X(S_D)$ produced by the Brownian paths at their first exit from a bounded domain D. Here the distribution of $X(L_{\bar{D}})$ produced by the same paths at their *last* exit from $\bar{D}$ yields the equilibrium measure for $\bar{D}$ through (11.16) and (11.17). Indeed, the last exit distribution $L_{\bar{D}}(x, \cdot)$ tells a lot more than $\mu_{\bar{D}}$, owing to the additional variable x there. Both measures H_D and $L_{\bar{D}}$ are supported by ∂D, namely the electrical charges stay on the surface, *because* Brownian paths are continuous!

Historically, it is curious that it took professional probabilists nearly three decades to take the little leap from a first exit (Kakutani 1944) to a last exit (1973).[24] To quote from [Chung P]:

> For some reason the notion of a last exit time, which is manifestly involved in the arguments, would not be dealt with openly and directly. This may be partially due to the fact that such a time is not an "optional" (or "stopping") time, does not belong to the standard equipment, and so must be evaded at all costs. — "Like the devil", as I added in a postcard to Henry McKean.

To wind up the story in a classic mood (mode), I will tell a little about what Gauss did. It is a physical principle that

Nature takes the least action and spends the least energy. The energy of two charges of amounts (positive or negative) λ_1 and λ_2 at the points x_1 and x_2 is given by the Newton–Coulomb law to be

$$\frac{\lambda_1 \lambda_2}{|x_1 - x_2|} .$$

A fantastic generalization (the circumspect Kellogg [p. 80] cautions about its validity) is as follows. For any two signed finite measures λ_1 and λ_2, their mutual energy is defined by

$$\langle \lambda_1, \lambda_2 \rangle = \iint \lambda_1(dx) g(x, y) \lambda_2(dy) \qquad (12.5)$$

provided the integral exists in the usual sense, after the decomposition into four parts using Hahn–Jordan's decomposition: $\lambda = \lambda^+ - \lambda^-$. We say λ has finite energy iff $\langle |\lambda|, |\lambda| \rangle < \infty$. It is a remarkable result that in the latter case $\langle \lambda, \lambda \rangle \geq 0$, and moreover $\langle \lambda, \lambda \rangle = 0$ only if λ is the zero measure. See [L; Sec. 5.2] for proofs, but they are not necessary for an appreciation of the rest of this discussion.

Gauss is known to be a terse person (although he gave seven proofs to a favorite theorem of his), and so used a more economic form of the energy above, to be denoted by

$$Q(\lambda) = \langle \lambda, \lambda \rangle - 2\lambda(1) . \qquad (12.6)$$

For any λ, we have by Fubini–Tonelli:

$$\langle \lambda^{\pm}, \mu_A \rangle = \int \lambda^{\pm}(dx) \int g(x, y) \mu_A(dy) = \lambda^{\pm}(1) . \qquad (12.7)$$

In particular

$$\langle \mu_A, \mu_A \rangle = \mu_A(1) < \infty . \qquad (12.8)$$

Moreover if λ has finite energy then so does $\lambda - \mu_A$.

Now consider an arbitrary signed finite measure λ having finite energy, and put $\lambda - \mu_A = \nu$. We have then

$$Q(\lambda) = \langle \mu_A + \nu, \mu_A + \nu \rangle - 2(\mu_A + \nu)(1)$$

$$= \langle \mu_A, \mu_A \rangle + 2\langle \mu_A, \nu \rangle + \langle \nu, \nu \rangle - 2\mu_A(1) - 2\nu(1)$$

$$= Q(\mu_A) + \langle \nu, \nu \rangle$$

by using (12.7) with ν for $\lambda^{\pm}$. Therefore, $Q(\lambda) \geq Q(\mu_A)$ $(= -\mu_A(1))$. Namely Q is minimized by μ_A among all the λ considered. This is Gauss's method of determining the equilibrium measure as achieving minimum energy according to Nature's predilection. However, he forgot to demonstrate the existence of such a minimizing measure, which lacuna was not filled until 1935 by Frostman. The same lacuna led to Dirichlet's Principle. As we saw here the Brownian motion paths know how to produce the equilibrium measure without bothering with energy. Nature's Ways are manifold. *Basta l'ultima uscita!*

Addenda

(I). Convergence at boundary

In the discussion of (7.5), I lamented the apparent lack of an expedient mode of convergence. Soon after the first edition of the book appeared, Giorgio Letta showed me such a mode by a brilliant detour. It is based on the fundamental theorem: a sequence of random variables convergent in measure contains a subsequence convergent almost surely [C; Third Edition; Theorem 4.2.3].

Following Letta, we begin by changing our notation for the Brownian motion process by specifying $X(0, w) = 0$ for all $w \in \Omega$, namely that all paths start at the origin in R^d, so that $x + X(.)$ is the BMP starting at $x \in R^d$. Now define the optional time, for a domain D:

$$S(x) = \inf\{t > 0 : x + X(t) \in D^c\}.$$

The distribution of $S(x)$ under P^0 is that of the previous S_D in (5.1) under P^x. We write P for P^0 and E for E^0 below. If $x \in \bar{D}$ then $x + X(S(x)) \in \partial D$. In our new notation, (7.6) becomes

$$\lim_{x \to z} E\{S(x)\} = 0.$$

Thus the random variable $S(x)$ converges to 0 in $L^1(P)$, *a fortiori* in P-measure. Hence any sequence x_n convergent to z contains a subsequence y_n such that $S(y_n)$ converges to 0 a.s.

The latter "strong" mode of convergence carries with it the a.s. convergence of $X(S(y_n))$ to $X(0)$, because $t \to X(t, w)$ is continuous for all or almost all w. Then of course $y_n + X(S(y_n))$ converges a.s. to $z + X(0) = z$. It follows by Lebesgue's bounded convergence theorem that for any bounded continuous function f on ∂D we have

$$\lim_n E\{f(y_n + X(S(y_n)))\} = f(z).$$

For the sake of clarity, put $c(x) = E\{f(x + X(S(x)))\}$. Looking back we see that we have proved: any sequence $c(x_n)$, where $x_n \to z$, contains a subsequence which converges to $f(z)$. This implies $c(x) \to f(z)$ as ($x \to z$, by a basic property of real numbers which we will call the "Sotto-Sotto" Lemma; See [C; p.89]. In our previous notation $c(x) = E^x\{f(X(S_D))\}$; thus (7.16) is proved.

Exercise for the learner: Prove that $X(S(x))$ converges in P-measure to 0 as $x \to z$. Beware: there are two hidden w there, see warning on p. 44.

(II). Time arrow and Green Symmetry

In Note 24 and elsewhere, reversing the time is hinted by "first entrance" versus "last exit". There is a junior case of this: in a fixed interval of $[0, t]$ reversing $X(s)$ to $X(t - s)$ for all s. Let

$$P^* = \int_{R^d} dx P^x,$$

which means assigning the nonprobability Lebesgue measure to $X(0)$ as initial "distribution". *Pronto* we have, for any two Borel sets A and B,

$$P^*\{X(0) \in A; X(t) \in B\} = P^*\{X(t) \in A; X(0) \in B\}. \quad (*)$$

This is proved by explicit calculation based on the symmetry of the transition density $p(.; x, y)$. When $A = R^d$, we see that $X(t)$ has the Lebesgue distribution for all t, namely the Lebesgue measure is invariant under the semigroup P_t. Now consider a cylinder set as in (6.7):

$$\Lambda = \bigcap_{j=1}^{n} \{X(t_j) \in A_j\}$$

where $0 = t_0 < t_1 < t_2 < \cdots < t_n = t$ and the A_j's are Borel sets with $A_0 = A$, $A_n = B$. Denote $\tilde{\Lambda}$ the "reversed" set in which each t_j becomes $t - t_j$. Then the relation (*) may be enhanced by inserting Λ and $\tilde{\Lambda}$ respectively into the left and right members. The proof is easy once the intuitive meaning of the symmetry of the transition is grasped, whereas an explicit check by multiple integral (as shown on p.184 of [L] and p.35 of [F]) may be hard to read! Next, let the relation be steadily refined by increasing the set of partition points t_j to a denumerable dense set in $(0, t)$. Then $\{t - t_j\}$ is likewise dense. Now let D be a compact set and take all the A_j to be D, except A and B to be arbitrary Borel subsets of D. The continuity of $X(.)$ implies then

$$P^*\{X(0) \in A; X(s) \in D \text{ for } 0 < s < t; X(t) \in B\}$$
$$= P^*\{X(t) \in A; X(s) \in D \text{ for } 0 < s < t; X(0) \in B\}.$$

Finally let D be an open set and let D_n be compacts increasing strictly to D. Replacing D by D_n in the above and letting $n \to \infty$, we conclude that the preceding relation remains true for an open D, and now (why only now?) may be recorded using our previous notation in (5.1) as

$$P^*\{X(0) \in A; t < S_D; X(t) \in B\}$$
$$= P^*\{X(0) \in B; t < S_D; X(t) \in A\}. \qquad (**)$$

Introduce the kernel

$$P_t^D(x, A) = P^x\{t < S_D; X(t) \in A\}$$

so that the Green potential in (6.12) may be written as

$$G_D = \int_0^\infty P_t^D\, dt.$$

In the new symbolism (**) becomes

$$\int_A dx.P_t^D(x, B) = \int_B dx.P_t^D(x, A). \qquad (***)$$

Integrating over t and invoking Fubini we obtain

$$\int_A dx.G_D(x, B) = \int_B dx.G_D(x, A).$$

But we know G_D has a nice density that is Green's (sic) function, given in (8.9) and is obviously (extended) continuous. Hence the above reduces to the double integrals of Green's function $g_D(x, y)$ over A and B in one way and over B and A in the other. Anybody can deduce from this that it is symmetric in (x, y).

The above is due to Z.Q. Chen to whom I proposed the "problem". In fact, the relation (***) would imply the symmetry of Gil Hunt's "relative" (his terminology) transition density $p^D(.; x, y)$, if only we knew the existence and continuity of the latter, as Einstein found out by solving the heat equation in the non-relative case of a "free" Brownian motion, see Sec. 3. In Hunt's proof, reproduced faithfully in [F; 2.2], all is done together but the work is rather tedious. A lot of TIME has flown by; is there now an easier way? Merely solving Einstein's equation with a boundary constraint won't do (why?).

It is legend that the time arrow is reversible in the great equations of physics from Newton to Maxwell. It is highly plausible that this has to do with certain "duality" phenomenon in classical potential theory. But Hunt has to impose extra mathematical assumptions in dealing with them in his "Hunt process" (1957-8), later extended by Blumenthal and Getoor, and others. Since then a general theory of reversing Markov process has been constructed by Chung and Walsh (see [T] and subsequent publications), yet regretably few concrete, physically significant applications seem to have resulted. It is partly in the hope of stimulating further progress in turning the arrow that this Addendum (II) is given here.

NOTES

(1) All the mathematics in this section can be read in Green's Essay cited in the References. The identity (1.1) here is his equation (2). The representation (1.9) is his equation (6), of which he said: "This equation is remarkable on account of its simplicity and singularity, seeing that it gives the value of the potential for any point p' within the surface, when its value at the surface itself is known, together with the density that a unit of electricity concentrated in p' would induce on this surface, if it conducted electricity perfectly, and were put in communication with the earth." [Two dispensable symbols have been deleted from the original text, on p. 32.]

(2) In the symbolism of Laurent Schwartz's "distribution", the calculations below amount to proving that

$$\Delta \left(\frac{1}{|x|} \right) = -4\pi\delta_0 \quad \text{in } R^3$$

where δ_0 is the Dirac function. This corresponds to Green's equation (1) [*numero uno!*] on p. 22. When I mentioned it to Laurent, he said he had not known.

(3) Did any of the ancient greats (perhaps Poincaré, who knew both potential and probability) suspect a true bond between Newton–Coulomb and DeMoivre–Laplace–Gauss? Could it be Euler who first evaluated the so-called infinite or

improper integral in (2.3)? I asked the late André Weil, who answered that he knew only Euler's number theory, but the question seemed interesting to him.

(4) In my book [E] I told the story of my rencontre with Dr. Einstein on Mercer Street, and his view of the foundation of probability. He also questioned me on the civil war in China, and opined that the United States should not get involved in it.

(5) Paul Lévy in his book [P] attributed to L. Bachelier's *Thèse: Théorie de la spéculation* (1900) as the first mathematical work on Brownian motion by probabilistic methods. Feller in his book (Volume 2) even called it the "Bachelier–Wiener" process. Bachelier made a passionate plea for probabilities in continuous time and with absolutely continuous (distribution) functions, no approximations! As an application he mentioned (p. 333ff.) the analytic theory of heat, wrote down the normal density as solution to Fourier's (heat) equation. There is no question that his "new method" (title of one of his books), for which he claimed sole credit, was applicable to Brownian motion which should be known to him through (the French) Jean Perrin if not (the foreigner) Albert Einstein. Indeed, the movement of material particles due to *hasard* was treated in two chapters (XX, XXI) under the titles of "kinetic" and "dynamic" probabilities, although his principal model was the stock market. [His master Poincaré, whose patronage he acknowledged, was known to frequent roulette tables.] But after spending many hours leafing through his *Thèse* and his grand *Calcul, tome* 1 (is there a tome 2?), and three paperback monographs dated 1937, 1938 and 1939 which I bought I do not remember when, I was unable to spot anywhere the name Brown (or *brownien*). However, Lévy stated [Q; p. 123] that Bachelier published a new book on Brownian motion just

before the war "of 1939–1945". [For me the war was at least of 1937–1945.] I have not found that book and have never seen it mentioned by anybody else, but I have written to Michel Weil at Besançon, where Bachelier had his post, to find out. Enfin, rien.

Personally I owe a great debt to Bachelier for a combinatorial formula in his *Calcul*, dealing with the Bernoullian case of $\max\limits_{1 \le k \le n} |S_k|$ in my thesis (1947). How did I know to look into that formidable volume on the fourth floor of the old Fine Hall and to find what I wanted, is a case of *la chance*. I looked at it again last night on pp. 252–253, but could no longer decipher it. [April 10, 1995.]

(6) Lévy [Q; p. 97]: "Since I have decided to conceal nothing of my own feelings, the moment has arrived to say that, among the opportunities that I had not been able to seize, that of leaving it to Wiener the discovery of this function $X(t)$, despite the fact that existence of precursor diminishes its importance, was one of those that I regretted most." The fairly literal translation is mine. Apropos I should also quote half a sentence from Wiener [D]: "The present paper owes its inception to a conversation which the author had with Professor Lévy in regard to ...".

(7) Years ago, a then young probabilist submitted a paper in which he announced, more or less, that the notion of a separable process was obsolete. This reminds me of many youngsters of today who had never dipped a pen in ink. In contemporary culture, such a fountain pen (called also ink-pen) beautifully crafted and gilded, is still in vogue as a gift item.

(8) In 195?, at a seminar in Columbia University, I showed something using the Markov property. A colleague Areyh Dvoretzky stood up after my talk, went to the blackboard and gave a shorter argument based on the mis-use of the property

as indicated above. Of course he admitted the faux-pas on the spot and no harm resulted. According to Feller, the same mistake occurred in a popular expository article by a renowned scientist, but since I was unable to retrace it in print, it is better not to spread the gossip.

(9) This little exercise was assigned to the audience (a kind of summer institute) in Dalian, China, 1991. Later I asked the local instructor to spread it among students and staff. I regret to report that after two years nobody had done the problem. Is it really that hard?

(10) I consulted Jean-Claude Zambrini on Einstein's text. He responded: "Einstein needed the Markovian character of his Brownian particle and not the independence of increments over intervals τ in a modern probabilistic sense." A more "realistic" model has been proposed by L. S. Ornstein (already mentioned by Fürth) and Uhlenbeck, but that is another story.

(11) Green used "potential function", Gauss used "potential"; see Kellogg [p. 52].

(12) I owe O. Kallenberg this attribution; previously I attributed it to E. Dynkin.

(13) The *bounded* measurability of $H_D f$ is essential here. Suppose h is merely finite and integrable over every sphere in R^d ($d \geq 1$) and the sphere-averaging property holds for all x and all $B(x, r)$. Is then h harmonic in R^d? Nobody seems to know the answer but I should conjecture "no", except for $d = 1$ where the answer is known to be yes. Local integrability, namely the integrability of h over every ball, is a sufficient supplementary condition for h to be harmonic. This is a wonderful case of the non-use and mis-use of the Fubini–Tonelli theorem

reviewed in Sec. 2 above. In an early draft of [F], Z. Zhao made the mistake of deriving (6.14) from (6.13) "simply by integrating". My treatment in [L; pp. 154–6] is more cautious, but some just don't get it.

(14) In general Markov processes, the point x is defined to be regular for the set A iff $E^x\{T_A\} = 0$. Thus here z is regular for D^c.

(15) In F. Knight's book *Brownian Motion And Diffusion* (1981), the proof of this assertion is by these words: "By the boundedness and continuity of f" (p. 67). He might as well have said, "Since two and two is four". I know Frank well and he is an honorable man. Was he aware of the little difficulty there or was he merely pretending to be concise? I propose to any instructor of this stuff to set a quiz for his class to obtain an opinion poll on the problem. Since the editor of the Mathematical Surveys originally wanted me to review Frank's manuscript but I declined, this is a belated comment for the edification of future authors.

(16) He proved that in his differential space the set of $x(t)$ in the sphere of radius r and satisfying the Hölder condition

$$|x(t_2) - x(t_1)| > ar(t_2 - t_1)^{1/2 - \varepsilon}, \quad 0 < \varepsilon < \frac{1}{2},$$

"for rational t, be it understood" (he warned), has an outer measure in Daniell's sense which goes to zero as a goes to infinity.

(17) I consulted a physicist (gravity expert) on this question, and was referred to Maxwell's book. Later I found out from the Appendix to [Green, p. 327] the following statement. "The important theorem of reciprocity, established in Art. 6, may be put in a clearer light by the following demonstration,

which is due to Professor Maxwell." It seems a little odd (to me) that neither he nor Maxwell used the argument indicated here.

I should add that Hunt in his 1958 paper has gone deeper into this symmetry. He constructed the transition density of the Brownian motion "killed off D", denoted by $p^D(t; x, y)$, so that for any Borel set A:

$$\int_A p^D(t; x, y) dy = P^x\{t < S_D;\ X_t \in A\} \ .$$

Then the Green function for D can be *defined* simply by

$$g_D(x, y) = \int_0^\infty p^D(t; x, y) dt$$

just as $g(x, y)$ can be defined by (2.3). He proved that for each $t > 0$, $p^D(t; x, y)$ is symmetric in $(x, y) \in D \times D$. The symmetry of $g_D(x, y)$ becomes a trivial consequence. By the way, there is NO probabilistic intuition for the symmetry of $p^D(t; \cdot, \cdot)$ although Hunt tried to make it appear a little more plausible in his argument. In [F] we give a straightforward computational *verification*.

(18) An apparently more intuitive way to define the pre-T tribe is the tribe generated by the stopped process $X(t \wedge T)$, $t \in R_0$. But this is inadequate because the random time T does not belong to it, as pointed out in [O]. In drafting this manuscript I found in Hunt's paper [p. 299] that he was quite aware of this, unlike another author who had adopted the defective definition without reflection. The present simple definition is due to A. Yushkevich. There is another longer but pleasant alternative: for a completely arbitrary random time T

we can define a strict-pre-T tribe denoted by $\mathcal{F}_{T-}$, then define $\mathcal{F}_{T+}$ to be $\bigwedge\limits_{n=1}^{\infty} \mathcal{F}_{(T+\frac{1}{n})-}$. What could be more natural? Now if T is optional, *mirabile dictu*, the latter is exactly the $\mathcal{F}_{T+}$ defined in the text. I'll let you figure out how to define that $\mathcal{F}_{T-}$ but you may consult [L; Sec. 1.3 and Exercise 4]. By the way, to appreciate a "trivial thing" like $\mathcal{F}_{T+}$, read Doob's 1953 book to see how he messed it up there.

(19) Regarding the origin of the strong Markov property, Hunt said [p. 294]: "Although mathematicians use this extended Markov property, at least as a heuristic principle, I have nowhere found it discussed with rigor. We begin by proving a precise version for Brownian motion .... I have not pushed the scope of the proof to the limit because it requires continuity of the sample functions on the right and it is probable* that a version of the extended Markoff property holds for processes which are only separable." In the case of Markov chains, namely on a countable state space, the said property is proved by me using the version of the process whose sample functions are lower-semi-continuous on the right, which is deduced from their separability [M; Sec. II.9]. In general state space with right continuous sample function and a Feller condition, the property was proved by R. M. Blumenthal and A. Yushkevich, independently; see e.g., [L; Sec. 2.3]. The historical account given in a recent pamphlet is inaccurate, according to the last-named author, but I have not seen it.

(20) As many popular books by and about Richard Feynman have appeared, allow me to add an anecdote here which I had told at a conference at Cornell University in 1993. One summer night, "when we were all young", I met Richard at

---

*Gil's surmise is wrong.

the popular night-spot College Spa on State Street in Ithaca, now defunct. One thing or another made me propose a little mathematical quiz to him: let an arbitrary triangle be given, mark a third length on each side and draw a line segment joining the division point to the opposite vertex. Lo and behold a smaller triangle appears in the middle, being one of the seven into which original triangle has been partitioned. Prove that its area is exactly one seventh of the original! Richard found this unbelievable, took out his pen or pencil, started computing on a napkin grabbed from the table, and after a couple of minutes announced that it must be wrong because his approximation did not support it. I bet him a quarter, and he re-did his computations. Eventually he paid up gamely after I told him to check it for an equilateral triangle.* My old friend Carlo Riparbelli (an aeronautical engineer) was with us, and many years later sent me two different proofs on large drafts-man's sheets. Shortly before Feynman's death I telephoned him (to my surprise he had a listed number) to ask his permission to quote him on the "eternal futility" (my words, but he said so in substance) not only of high mathematics but also high physics and astronomy, spoken by him on the Nova program and available in print ("The pleasure of finding things out", January 25, 1983). My article containing his remarks was published in Singapore, 1984, after I gave a talk there. Later, the late Mr. Shi Hóng-gao, who was my English teacher in high school and held ministerial positions for Mao's China in the fifties–sixties, translated it into Chinese because, as he told me, it should serve as a "cooling medicine" for the current

---

*Such a quick solution by projection was found by my other friend Armin Perry, who took my Advanced Calculus at Cornell and later worked at General Electric at Syracuse when I told him the problem. He did it without any hint.

Chinese science scene. Despite his good connections his manuscript was refused publication as being counter-something-or-
other, until 1988 shortly before the Tiananmen tragedy (*Nature
Journal*, vol. 11, no. 9, 1988, Shanghai). The editor Mr. Zhang
was dismissed after the tragedy as too "liberal" — I heard.
Even Feynman, in the committee investigating the fatally
exploded rocket (1986), had to buck the politicians and bureau-
crats to present his testimony with a glass of (icy?) water to
the U.S. congress and television audience. It was a marvellous
performance.

(21) After I proposed this little problem at the Fifteenth
Seminar On Stochastic Processes held at the University of
Florida in Gainesville, March 9–11, 1995 and got no response,
I asked my colleague Y. Katznelson. He wrote down an example with explanation in six lines (which may need a little
amplification). It is one of those things in mathematics where
if you *really* see the pinch you can fix the shoe. Try it.

(22) The adjective "left" should sound a drum here. A
Hunt process has right-continuous paths, and an optional time
is approximated from the right, a crucial step in the derivation
of the strong Markov property from the (simple, unqualified)
Markov property under certain conditions, including the case
of Brownian motion (not given in this book but see Note 19).
But when the last exit is concerned, time must be reversed
and so "right" becomes "left". You will see presently that
approximating L_A from the right will not do.

(23) When A is the unit ball one can compute μ_A directly
from (11.7) and (11.14). I asked Z.Q. Chen to do it. He
used spherical coordinates and confirmed that it is the uniform distribution over the surface. The details are a bit too
long to be given here. Sometimes an abstract general argument

does better than a special calculation — as if we did not know that!

(24) Whereas the notion of a first happening was utilised early in probability arguments, that of a last happening was overlooked. As far as I know, "last exit" made its first formal entry around 1950 in a decomposition formula in Markov chains, as dual to "first entrance"; see my elementary book [E; Sec. 8.4] and the reminiscent [R]. In continuous time the notion is much more subtle and requires careful handling. Of course, if the legendary time-arrow is reversed in direction, then future becomes past and first becomes last. Easy to say but hard to do, rigorously. Physicists talk a lot about reversibility and some of the electrical phenomena discussed in this book have been interpreted in that light. For example, the equilibrium problem has an extension from one conductor to two (or more), known as the "condenser problem". This can be treated by probabilistic methods, but we (Chung and Getoor, 1977) found that certain "duality hypothesis" was needed. We know (see [T]) that any decent Markov process can be reversed in time to yield another (slightly weaker) one, but unfortunately this innate reversibility has not produced the analytic dual structure suitable for classical potential theory, see e.g. [L; Chapter 5]. In a sense, the consideration of last exit serves as an easier way out of the intractable time-reversal.

REFERENCES

Bachelier, L.
   Calcul des probabilités, Tome 1, Gauthier-Villars, 1912.

Chung, Kai Lai
   [C] *A Course in Probability Theory*, Third Edition, Academic Press, 2001 [First Edition 1968].
   [D] Doubly-Feller process with multiplicative functional, *Seminar on Stochastic Processes*, Birkhäuser, 1985.
   [E] *Elementary Probability Theory with Stochastic Processes*, Third Edition, Springer-Verlag, 1978 [First Edition 1974].
   [G] Greenian bounds for Markov processes, *Potential Analysis* **1**, 1992, pp. 83–92.
   [I] *Introduction to Random Time*, Monographs of the Portugese Mathematical Society, No. 1, 2001.
   [L] *Lectures from Markov Processes to Brownian Motion*, Springer-Verlag, 1982.
   [M] *Markov Chains with Stationary Transition Probabilities*, Second Edition, Springer-Verlag, 1967 [First Edition 1960].
   [P] Probabilistic approach to the equilibrium problem in potential theory, *Ann. Inst. Fourier* **23**, 1973, pp. 313–322.

[R] Reminiscences of one of Doeblin's papers, *Contemporary Mathematics* **149**, 1993, pp. 65–71.

Chung, Kai Lai and Doob, J. L.
[O] Fields, optionality and measurability, *Amer. J. Math.* **87**, 1965, pp. 397–424.

Chung, Kai Lai and John B. Walsh
[T] *To Reverse a Markov Process Acta Math.* **123**, 1969, pp. 225–251.

Chung, Kai Lai and Zhao, Z.
[F] *From Brownian Motion to Schrödinger's Equation*, Springer-Verlag, 1995.

Courant, R. and Hilbert, D.
Methoden der mathematischen Physik I, Springer, Berlin, 1937.

Einstein, Albert
Investigation on the Theory of Brownian Movement, Dover 1956 [Translation originally published in 1926].

Feller, William
An Introduction to Probability Theory and its Applications, Volume 2, Second Edition, Wiley, 1971 [First Edition 1966].

Green, George
An essay on the applications of analysis to the theories of electricity and magnetism, Nottingham, 1928. [In the collection: Mathematical Papers, originally published in 1870, reprinted by Chelsea 1970.]

Hunt, G. A.
Some theorems concerning Brownian motion, *Trans. Amer. Math. Soc.* **81**, 1956, pp. 294–319.

Kakutani, S.
   Two-dimensional Brownian motion and harmonic functions, *Proc. Imp. Acad. Tokyo* **22**, 1944, pp. 706–714.

Kellogg, O. D.
   Foundations of Potential Theory, Springer-Verlag, 1929.

Lévy, Paul
   [P] *Processus stochastiques et mouvement brownien*, Second Edition, Gauthier-Villars, 1956 [First Edition 1948].
   [Q] *Quelques aspects de la pensées d'un mathématicien*, Albert Blanchard, 1970.

Royden, H.
   Real Analysis, Second Edition, Macmillan, New York, 1968.

Titchmarsh, E. C.
   The Theory of Functions, Second Edition, Oxford, 1939 [First Edition 1932].

Wiener, Norbert
   [D] Differential space, *J. Math. Phys.* **2**, 1923, pp. 131–174.
   [T] The Dirichlet problem, *J. Math. Phys.* **3**, 1924, pp. 127–147.

SOME CHRONOLOGY

Newton	1642–1727
DeMoivre	1667–1754
Euler	1707–1783
Coulomb	1736–1806
Laplace	1749–1827
Brown	1773–1858
Gauss	1777–1855
Poisson	1781–1840
Cauchy	1789–1857
Faraday	1791–1867
Green	1793–1841
Dirichlet	1805–1859
Weierstrass	1815–1897
Maxwell	1831–1879
Poincaré	1854–1912
Markov	1856–1922
Borel	1871–1956
Lebesgue	1875–1941
Einstein	1879–1955
Lévy	1886–1974
Wiener	1894–1964
Kolmogorov	1903–1987
Feller	1906–1970
Feynman	1918–1988

APPENDICES

(I) George Green

INTRODUCTORY OBSERVATIONS.

THE object of this Essay is to submit to Mathematical Analysis the phenomena of the equilibrium of the Electric and Magnetic Fluids, and to lay down some general principles equally applicable to perfect and imperfect conductors; but, before entering upon the calculus, it may not be amiss to give a general idea of the method that has enabled us to arrive at results, remarkable for their simplicity and generality, which it would be very difficult if not impossible to demonstrate in the ordinary way.

It is well known, that nearly all the attractive and repulsive forces existing in nature are such, that if we consider any material point p, the effect, in a given direction, of all the forces acting upon that point, arising from any system of bodies S under consideration, will be expressed by a partial differential of a certain function of the co-ordinates which serve to define the point's position in space. The consideration of this function is of great importance in many inquiries, and probably there are none in which its utility is more marked than in those about to engage our attention. In the sequel we shall often have occasion to speak of this function, and will therefore, for abridgement, call it the potential function arising from the system S. If p be a particle of positive electricity under the influence of forces arising from any electrified body, the function in question, as is well known, will be obtained by dividing the quantity of electricity in each element of the body, by its distance from the particle p, and taking the total sum of these quotients for the whole body, the quantities of electricity in those elements which are negatively electrified, being regarded as negative.

111

(II) L. Bachelier

PRÉFACE.

Cet Ouvrage n'a pas seulement pour but d'exposer, avec quelques développements, l'ensemble des connaissances acquises dans le passé sur le calcul des probabilités, il a aussi pour objet de faire connaître de nouvelles méthodes et de nouveaux résultats qui représentent, à certains points de vue, une transformation complète de ce calcul.

La conception des probabilités continues constitue la base de ces nouvelles études. On pensait précédemment que seules des formules discontinues pouvaient être des conséquences exactes des principes du calcul des probabilités, et cette idée était d'autant plus naturelle que les problèmes traités alors ne pouvaient admettre d'autres genres de solutions.

On employait bien parfois des formules continues mais on les considérait comme approchées, de sorte qu'elles ne pouvaient servir de base pour de nouvelles recherches. Pour cette raison, leur emploi ne s'est pas généralisé depuis Laplace.

L'idée de considérer les probabilités comme continues fut seulement envisagée il y a quelques années lorsque je me proposai de résoudre des questions ne pouvant admettre que des solutions continues exactes.

La théorie édifiée alors était assez particulière, des ébauches publiées dans différents Recueils indiquent le cours de son évolution; généralisée successivement dans tous les sens, elle a pris un tel développement qu'il m'a semblé nécessaire de la

(III) Wiener

DIFFERENTIAL-SPACE
By Norbert Wiener

§1. Introduction. The notion of a function or a curve as an element in a space of an infinitude of dimensions is familiar to all mathematicians, and has been since the early work of Volterra on functions of lines. It is worthy of note, however, that the physicist is equally concerned with systems the dimensionality of which, if not infinite, is so large that it invites the use of limit-processes in which it is treated as infinite. These systems are the systems of statistical mechanics, and the fact that we treat their dimensionality as infinite is witnessed by our continual employment of such asymptotic formulae as that of Stirling or the Gaussian probability-distribution.

The physicist has often occasion to consider quantities which are of the nature of functions with arguments ranging over such a space of infinitely many dimensions. The density of a gas, or one of its velocity-components at a point, considered as depending on the coördinates and velocities of its molecules, are cases in point. He therefore is implicitly, if not explicitly, studying the theory of functionals. Moreover, he generally replaces any of these functionals by some kind of average value, which is essen-

(IV) Kakutani

143. Two-dimensional Brownian Motion and Harmonic Functions.

By Shizuo KAKUTANI.

Mathematical Institute, Osaka Imperial University.

(Comm. by T. TAKAGI, M.I.A., Dec. 12, 1944.)

1. The purpose of this paper is to investigate the properties of two-dimensional Brownian motions[1] and to apply the results thus obtained to the theory of harmonic functions in the Gaussian plane. Our starting point is the following theorem : *Let D be a domain in the Gaussian plane R^2, and let E be a closed set on the boundary Bd(D) of D. Then, under certain assumptions on D and E, the probability $P(\zeta, E, D)$, that the Brownian motion starting from a point $\zeta \in D$ will enter into E without entering into the other part Bd(D) − E of the boundary of D before it, is equal to the harmonic measure in the sense of R. Nevanlinna of E with respect to the domain D and the point ζ.*

It is expected that, by means of this method, many of the known results in the theory of harmonic or analytic functions will be interpreted from the standpoint of the theory of probability. We shall here give only the fundamental results and a few of its applications, leaving the detailed discussions of further applications to another occasion.

Most of the results obtained in this paper are also valid for the case of higher dimensional Brownian motions. But there are also many theorems in which the dimension number plays an essential rôle[2]. For example, Theorems 6, 7 and 8 of this paper are no longer true in R^3. The situation will become clearer if we observe the following theorem : *Consider the n-dimensional Brownian motion in $R^n (n \geq 2)$, and let $\bar{K}^n$ be the closed unit sphere in R^n. Then, for any $\zeta \in R^n - \bar{K}^n$, the probability $P(\zeta, \bar{K}^n, R^n - \bar{K}^n)$ that the Brownian motion starting from ζ will enter into $\bar{K}^n$ for some $t > 0$ is equal to $|\zeta|^{2-n}$ if $n \geq 3$[3], while this probability is $\equiv 1$ on $R^n - \bar{K}^n$ if $n = 2$[5].* This result is closely related with the fact that there is no bounded harmonic function, other than the constant 1, which is defined on $R^n - \bar{K}^n$ and tends to 1 as $|\zeta| \to 1$, while, for any $n \geq 3$, $u(\zeta) = |\zeta|^{2-n}$ is a non-trivial example of a bounded harmonic function with the said property.

2. Let $\{z(t, \omega) = (x(t, \omega), y(t, \omega)) \mid -\infty < t < \infty, \omega \in \Omega\}$ be a two-dimensional Brownian motion defined on the $z = (x, y)$-plane R^2, i. e. an independent system of two one-dimensional Brownian motions $\{x(t, \omega) \mid$

1) Brownian motions were discussed by N. Wiener and P. Lévy. Cf. N. Wiener, Generalized harmonic analysis, Acta Math., **54** (1930); N. Wiener, Homogeneous chaos. Amer. Journ. of Math., **60** (1939); R. E. A. C. Paley and N. Wiener, Fourier transforms in the complex domain, New York, 1933; P. Lévy, L'addition des variables aléatoires, Paris, 1937; P. Lévy, Sur certains processus stochastiques homogènes, Compositio Math., **7** (1939); P. Lévy, Le mouvement brownien plan, Amer. Journ. of Math., **61** (1940).

2) Cf. R. Nevanlinna, Eindeutige analytische Funktionen, Berlin, 1937.

3) Cf. S. Kakutani, On Brownian motions in n-space, Proc. **20** (1944).

4) $|\zeta|$ denotes the Euclidean distance of ζ from the origin of R^n.

5) The case $n = 2$ is contained in Theorem 4 of this paper.

(V) Hunt

SOME THEOREMS CONCERNING BROWNIAN MOTION

BY

G. A. HUNT[1]

Given a random time T and a Markoff process $X(\tau)$ with stationary transition probabilities, one can define a new process $X'(\tau) = X(\tau + T)$. It is plausible, if T depends only on the $X(\tau)$ for τ less than T, that $X'(\tau)$ is a Markoff process with the same transition probabilities as $X(\tau)$ and that, when the value of $X'(0) \equiv X(T)$ is fixed, the process $X'(\tau)$ is independent of the history of $X(\tau)$ before time T. Although mathematicians use this extended Markoff property, at least as a heuristic principle, I have nowhere found it discussed with rigor. We begin by proving a precise version for Brownian motion (Theorem 2.5, with an extension in §3.3). Our statement has the good points that the hypotheses are easy to verify, that the proof is thoroughly elementary (it even avoids conditional probabilities), and that it holds for all processes with stationary independent increments (see §3.1). I have not pushed the scope of the proof to the limit because it requires continuity of the sample functions on the right and it is probable that a version of the extended Markoff property holds for processes which are only separable.

In §4 and §5 we use the extended Markoff property to study the transition probability density $q(\tau, r, s)$ of a Brownian motion in R^n in which a particle is extinguished the moment it hits a closed set E. It turns out that $q(\tau, r, s)$, which can be defined without ambiguity for every r and s in R^n, is the fundamental solution of the heat equation on $R^n - E$ with boundary values 0 (at the regular points of E), and that $G(r, s) = \int_0^\infty q(\tau, r, s) d\tau$ is the Green's function for the Laplacian on $R^n - E$. The content of these sections must be considered more or less as common knowledge; however, I believe some of the details are new, especially those concerning irregular points and the probabilistic interpretation.

In writing §4 and §5 I profited from many discussions with Mark Kac and Daniel Ray. Had I known of Doob's fundamental paper [4] on the heat equation it is likely that these sections would have been cast in a different form.

Kac has made the following conjecture. Consider in the plane a closed bounded domain E and a point r outside E; the probability that a Brownian motion starting from r does not meet E by time τ is for large τ asymptotic to $2\pi H(r)/\ln \tau$. Here $H(r)$ is the Green's function of $R^2 - E$ with pole at in-

---

Received by the editors July 20, 1955.

[1] This research was supported in part by the United States Air Force, through the Office of Scientific Research of the Air Research and Development Command.

(VI) Chung

PROBABILISTIC APPROACH
IN POTENTIAL THEORY
TO THE EQUILIBRIUM PROBLEM

by Kai Lai CHUNG [1]

The problem of equilibrium is the first problem for the ancient period of potential theory recounted by Brelot in his recent historical article [1]. The existence of an equilibrium measure for the Newtonian potential was affirmed by Gauss as the charge distribution on the surface of a conductor which minimizes the electrostatic energy. But it was Frostman who rigorously established the existence in his noted thesis (1935), and extended it to the case of M. Riesz potentials. Somewhat earlier, F. Riesz had given his well-known decomposition for a subharmonic function which is a closely related result. For further history and references to the literature see Brelot's article. From the viewpoint of probability theory, the equilibrium problem in the Newtonian·case takes the following from :

$$\underline{P}^x\{T_B < \infty\} = \int_{\partial B} u(x, y) \, \mu_B(dy). \tag{1}$$

Here the underlying process is a Brownian motion $\{X_t, t \geq 0\}$ in R^3 ; $\underline{P}^x\{\cdots\}$ denotes the probability (Wiener measure) when all paths issue from the point x ; B is a compact set (the conductor body) ; $T_B = T_B(\omega)$ is the hitting time of B by the path ω :

$$T_B(\omega) = \inf\{t > 0 \mid X_t(\omega) \in B\} ;$$

∂B is the boundary of B ; $u(x, y)$ is the associated potential density

$$u(x, y) = \frac{1}{2\pi |x - y|} ;$$

(1) Research supported by the Office of Scientific Research of the United States Air Force, under AFOSR Contract F44620-67-C-0049.

Part II.

Brownian Motion on the Line

Generalities

The triple (W, F, P) is a probability space: W is an abstract space of points w, F is a Borel tribe of subsets of W containing W as a member, P is a probability measure on F.

A classic example: $W = [0, 1]$, $F =$ the Borel sets in $[0, 1]$, $P =$ the Borel measure. Another example: $W = [0, 1]$, $F =$ the Lebesgue measurable sets in $[0, 1]$, $P =$ the Lebesgue measure. The latter has the property of "completeness" not shared by the former. Namely a subset of a set of Borel measure zero may not be Borel measurable, but a subset of a set of Lebesgue measure zero is always Lebesgue measurable (hence has Lebesgue measure zero). It is always possible to "complete" the probability space and so we may as well assume this: if $A \subset B$, $B \in F$ and $P(B) = 0$, then $A \in F$.

We define a stochastic process on the triple as a collection (space) of functions with domain D, parametrized by w in W:

$$X(w, \cdot) : t \to X(w, t), \quad t \in D \, ;$$

such that for each $t \in D$, $X(\cdot, t)$ is a random variable on the triple, namely a function of w with domain W and measurable with respect to the Borel tribe F. In the following we shall take the range to be the compactified real line $[-\infty, +\infty]$. The random variable $X(., t)$ will also be denoted by $X_t(\cdot)$ or X_t. Elementary knowledge of random variables is presupposed.

An older definition of stochastic process is $\{X_t, t \in D\}$. In this form it is a collection (family) of random variables parametrized by t in D. The two definitions are equivalent and serve different ends, as we shall see.

We have yet to specify the parameter set D. When D is the set of natural numbers the second definition of a process becomes a sequence of random variables studied in basic probability theory. Here we shall concentrate on the case $D = [0, \infty)$, the positive (sic) real numbers, and t will be called the "time". Thus the stochastic process becomes the chance evolution of a random real variable $X(\cdot, t)$ with the passing of TIME.

The random variables $\{X_s, 0 \leq s \leq t\}$ generate a Borel tribe to be denoted by F_t; as t increases this increases (i.e. does not decrease) to F_∞. The random variables $\{X_s, t < s < \infty\}$ generate a Borel tribe denoted by F_t'.

The process $\{X_t\}$ is called a Homogeneous Markov Process iff for each t and each $A \in F_t'$ we have

$$P\{A|F_t\} = P\{A|X_t\} \tag{1}$$

where the left member is the conditional probability of A relative to F_t and the right member the conditional probability relative to the tribe generated by the single X_t. It is essential to understand the exact meaning of the conditioning (which is not a common notion in analysis, classic or functional!) When the random variable X_t takes discrete values x we may write $P\{A|X_t = x\}$; in general this symbolism must be interpreted in a Radon–Nikodym derivative sense as we shall do occasionally; see Chung [3; Sec. 9.1] for a full discussion.

When the time parameter t is the natural numbers, the homogeneous Markov Property (1) has a facile extension called The Strong Markov Property, as follows. Let T be a positive finite random variable such that for each natural number n we have $\{w : T(w) = n\} \in F_n$. Such a T is called "optional". Let F_T denote the collection of sets A in F such that for each n we have $A \cap \{w : T(w) = n\} \in F_n$. Then F_T forms a Borel tribe, a subtribe of F. Let F_T' denote the tribe generated by the "post-T process $\{X_{T+t}, t \in D\}$, where $X_{T+t}(w) = X(w, T(w) + t)$. Then the strong Markov property consists in replacing the constant t in (1) by the random T everywhere, viz. for each $A \in F_T'$ we have on the set $\{T < \infty\}$:

$$P\{A|F_T\} = P\{A|X_T\}. \tag{2}$$

When the time parameter t is in $D = [0, \infty)$, the situation is more delicate. A positive random variable T is called "optional" iff for each $t \in D$, we have

$$\{w : T(w) < t\} \in F_t. \tag{3}$$

For an optional T we define F_T to be the Borel tribe of sets A in F such that for each $t \in D$ we have

$$A \cap \{w : T(w) < t\} \in F_t.$$

The assertion that F_T is a Borel tribe is proved by using the definition of T. Define the post-T tribe F_T' as before.

A positive constant t_0 such as 17 is a random variable. Is it optional? We must check it out against the definition and the answer is YES. Next, when T is the constant t, is the tribe F_T equal to F_t? Again we must check it out, and the answer is NO. In fact, when $T = t$ we have

$$F_T = F_{t+} = \bigwedge_{s>t} F_s$$

which may be larger than F_t. This is what I meant by "delicate". However by some permissible manipulation, we may

suppose that for every $t \in D$, we have $F_{t+} = F_t$; see [L; p. 61].*
Then it follows that the condition (3) for optionality may be
changed by replacing the "<" there with "$\leq$". All this stuff
may seem hair-splitting nuisance to a beginner or a practical
man in a hurry, but is required by relentless logic. Now we are
ready to state the Strong Markov Property.

The homogeneous Markov process $\{X_t, t \in D\}$ is said to
have the strong Markov property if and only if (1) remains
true when the constant t is replaced by an optional random T.
When T is a constant t, this reduces to (1), as we have seen.

For continuous time parameter t, this stronger property is
no longer a consequence of the plain Markov property given
in (1). It must be assumed (as we shall) or proved under additional hypothesis on the process.

Returning to our first definition of process, we call the function of $t \in D : t \to X(w, t)$, or more simply $X(w, \cdot)$, the sample
function corresponding to w. When its range is specified as
above, it is a function of the real variable t taking real values
or the two limiting values $+\infty$ and $-\infty$. This is the kind of
functions studied in calculus. It may be continuous or discon-
tinuous at each point t and may go to infinity somewhere. For
a given stochastic process, the nature of its sample functions
depends on the initial hypotheses on the laws of probability
governing the movement (evolution) of X_t as the time t pro-
gresses. This is expressed by a transition probability kernel as
follows. For t in D, x in $(-\infty, +\infty)$, and Borel set A, we put

$$P_t(x, A) = P^x\{X_t \in A\} = P\{X_{s+t} \in A | X_s = x\}, \quad s \in D.$$

$$(4)$$

As a linear operator, P_t is defined on the class of bounded Borel
functions f by

---

*See References in Part I.

$$P_t f(x) = \int\limits_{-\infty}^{+\infty} P_t(x, dy) f(y)$$

with the semigroup property $P_{s+t} = P_s P_t$. Observe that there is NO random element in the semigroup, and read [GBAP,* Sec. 4] for commentary.

For the Brownian Motion [Movement] On Line, we have

$$P_t(x, A) = \int\limits_{A} \frac{1}{\sqrt{2\pi t}} e^{-\frac{(y-x)^2}{2t}} dy \qquad (5)$$

can also be seen as a linear operator acting on $\mathbb{1}_A$

where the integrand is the probability density of the normal distribution with mean x and variance t; it will be denoted as $p_t(x, y)$ or $p(t; x, y)$. Read [GBAP, Sec. 2] for historical perspective on this particular case. It is proved that all the sample functions of this process may be taken to be everywhere continuous; and it is a homogeneous Markov process with the strong Markov property: a "strong Markov process". The proofs of these fundamental properties of the Brownian Motion may be found in [L].

We end these preliminaries with a discussion of a primary class of optional random variables. For a process taking values in R^d, the Euclidean space of dimension $d \geq 1$, and any Borel set there, define

$$T_A(w) = \inf\{t > 0 : X(w, t) \in A\} \qquad (6)$$

where the inf is $+\infty$ when the set is void. We shall call T_A the "hitting time of A". If we change the "$t > 0$" in (6) into "$t \geq 0$", the result is a different function of w called the "first entrance time in A". The difference can be tremendous, see

---

*Namely Part I in this edition.

[GBAP, p. 25] for an anecdote. Now if the time parameter D is the natural numbers, it is trivial to verify that T_A is optional. When $D = [0, \infty)$, conditions must be imposed on the process and on the set A to make T_A optional. [By the way, how do we know T_A is even a random variable?]

We prove that if all sample functions are continuous, and A is either an open set or a closed set, then T_A is an optional random variable.

Let A be open. Then we have for $t > 0$, omitting w below:

$$T_A(\omega) = \inf\{t > 0 : X(\omega, t) \in A\}$$

$$\{T_A < t\} = \{\text{there exists } s \text{ in } [0, t) \text{ such that } X_s \in A\}$$

$$= \{\text{there exists } r \text{ in } [0, t) \cap Q \text{ such that } X_r \in A\}$$

where Q is the set of all rational numbers, and the equation is due to the openness of A [it may be false for a closed A!], as well as the continuity of $X(\cdot)$. Hence

$$\{T_A < t\} = \bigcup_{r \in [0,t) \cap Q} \{X_r \in A\} \in F_t$$

because $X_r \in F_t$, A is Borel, and F_t is a Borel tribe.

Let B be closed. There exists a sequence of strictly decreasing open sets O_n such that $\bigcap_{n=1}^{\infty} O_n = B$. Then we have

$$\{T_B \leq t\}$$

$$= \bigcup_{k=1}^{\infty} \bigcap_{n=1}^{\infty} \{\text{there exists } r \text{ in } [t/k, t] \cap Q \text{ such that } X_r \in O_n\}.$$

It is a good exercise in elementary topology to prove the equation above, which is more than needed for the optionality of T_B (if we do not assume $F_{t+} = F_t$ as discussed earlier).

In the case $d = 1$, namely in the real line, if A is an open interval or closed interval, and the process starts outside A, it is geometrically obvious that hitting the bounded interval (a, b) or $[a, b]$ is the same as hitting one of the two end points a or b; hitting the interval (b, ∞) or $[b, \infty)$ is the same as hitting the singleton b. Thus T_A is reduced to $T_a \wedge T_b$ or T_b. What happens when the process starts inside the interval is obvious except when it starts at an end-point. This peculiar situation is tricky and will be discussed for the Brownian motion below.

Introduction

These lecture notes are based on a short course given at Stanford University in the fall of 1979.* The audience consists of graduate students from several departments and some faculty auditors. They are supposed to know the elements of Brownian motion including the continuity of paths and the strong Markov property, as well as elementary martingale theory. The emphasis is on methodology and some inculcation is intended.

In the following $\{X(t), t \geq 0\}$ is a standard Brownian motion on $R = (-\infty, +\infty)$.

1. Exit and Return

Let (a, b) be a finite interval and put

$$\tau = \tau_{(a,b)} = \inf \{t > 0 : X(t) \notin (a, b)\}. \tag{1.1}$$

This is called the [first] exit time from (a, b). It is optional. If $x \notin [a, b]$, then clearly

$$p^x\{\tau = 0\} = 1 \tag{1.2}$$

---

*Journal of Mathematical Research and Exposition **1** (1) (1981) 73–82 and **2** (3) (1982) 87–98.

by the continuity of paths. If $x = a$ or $x = b$, the matter is not so obvious because the inf in (1.1) is taken over $t > 0$, not $t \geq 0$. Starting at a, it is conceivable that the path might move into (a, b) without leaving it for some time. It is indeed true that (1.2) holds for $x = a$ and $x = b$ but the proof will be given later. Our immediate concern is: if the path starts at x, where $x \in (a, b)$, will it leave the interval eventually?

We have the crude inequality

$$\sup_{x \in (a,b)} P^x\{\tau > 1\} \leq \sup_{x \in (a,b)} P^x\{X(1) \in (a, b)\}. \qquad (1.3)$$

The number on the right side can be expressed in terms of a normal distribution but it is sufficient to see that it is strictly less than one. Denote it by δ. The next step is a basic argument using the Markovian character of the process. For any $x \in (a, b)$ and $n \geq 1$:

$$P^x\{\tau > n\} \leq E^x\{\tau > n - 1; P^{X(n-1)}[\tau > 1]\}$$

$$\leq P^x\{\tau > n - 1\} \cdot \delta. \qquad (1.4)$$

In the above we have adopted the notation commonly used in Markov processes. It symbolizes the argument that if the path has not left (a, b) at time $n - 1$, then wherever it may be at that time, the probability does not exceed δ that it will remain in (a, b) for another unit of time. If follows by induction on n that

$$P^x\{\tau > n\} \leq \delta^n. \qquad (1.5)$$

Since $\delta < 1$ we obtain $P^x\{\tau = \infty\} = 0$ by letting $n \to \infty$. This answers our question: the path will (almost surely) leave the interval eventually:

$$P^x\{\tau < \infty\} = 1. \qquad (1.6)$$

In fact the argument above yields more. For any ε such that $e^\varepsilon \delta < 1$, we have

$$E^x\{e^{\varepsilon\tau}\} \leq \sum_{n=1}^{\infty} e^{\varepsilon n} P^x\{n-1 < \tau \leq n\} \leq \sum_{n=1}^{\infty} e^{\varepsilon n} \delta^{n-1} < \infty.$$
$$(1.7)$$

In standard terminology this says that the random variable τ has a generating function $E^x\{e^{\theta\tau}\}$ which is finite for sufficiently small values of θ. In particular it has finite moments of all orders.

At the exit time τ, $X(\tau) = a$ or $X(\tau) = b$ by continuity. For any $x \in (-\infty, \infty)$, let us define

$$T_x = \inf\{t > 0 : X(t) = x\}. \qquad (1.8)$$

This is called the *hitting time* of the set $\{x\}$; it is optional. It is clear that

$$\tau_{(a,b)} = T_a \wedge T_b. \; = min\,(T_a \, \mathop{>} T_b) \quad (1.9)$$

If we keep b fixed while letting $a \to -\infty$, then

$$\lim_{a \to -\infty} \tau_{(a,b)} = T_b. \qquad (1.10)$$

It is worthwhile to convince oneself that (1.10) is not only intuitively obvious but logically correct, for a plausible argument along the same line might mislead from (1.6) and (1.10) to:

$$P^x\{T_b < \infty\} = 1. \qquad (1.11)$$

The result is true but the argument above is fallacious* and its refutation is left as Exercise 1 below. We proceed to a correct proof.

Proposition 1. *For every x and b in $(-\infty, \infty)$, (1.11) is true.*

---

*It may be reported here that an applied analyst fell for it when asked in my class.

Proof. For simplicity of notation we suppose $x = 0$ and $b = +1$. Let $a = -1$ in (1.7). Since the Brownian motion is symmetric with respect to the directions right and left on the line, it should be clear that at the exit from $(-1, +1)$, the path will be at either endpoint with probability $1/2$. If it is at $+1$, then it hits $+1$. If not, the path is at -1, and we consider its further movement until the exit from $(-3, +1)$. According to strong Markov property this portion of the path is like a new Brownian motion starting at -1, and is stochastically independent of the motion up to the hitting time of -1. Thus we may apply (1.6) again with $x = -1$, $a = -3$ and $b = +1$. Observing that -1 is the midpoint of $(-3, +1)$, we see that at the exit from the latter interval, the path will be at either endpoint with probability $1/2$. If it is at $+1$, then it hits $+1$. If not, the path is at -3, and we consider its further movement until the exit from $(-7, +1)$, and so on. After n such steps, if the path has not yet hit $+1$, then it will be at $-(2^n - 1)$ and the next interval for exit to be considered is $(-(2^{n+1} - 1), +1)$. These successive trials (attempts to hit $+1$) are independent, hence the probability that after n trials the path has not hit $+1$ is equal to $(\frac{1}{2})^n$. Therefore the probability is equal to $\lim_{n \to \infty} (\frac{1}{2})^n = 0$ that the path will never hit $+1$; in other words the probability is 1 that it will hit $+1$ eventually.

The scheme described above is exactly the celebrated gambling strategy called "doubling the ante". [The origin of the name "martingale".] The gambler who is betting on the outcome of tossing a fair coin begins by staking \$1 on the first outcome. If he wins he gets \$1 and quits. If he loses he flips the coin again but doubles the stake to \$2. If he wins then his net gain is \$1 and he quits. If he loses the second time he has lost a total of \$3. He then repeats the game but redoubles the stake to \$4, and so on. The mathematical theory above shows that if the game is played in this manner indefinitely, the gambler stands to gain \$1 sooner or later. Thus it is a "sure win" system, the only drawback being that one needs an infinite amount of money to play the system. To get a true feeling for the situation one should test the gambit by going to a roulette table and bet repeatedly on "red or black", which is the nearest thing to a fair game in a casino.

There are many other proofs of Proposition 1, of which one will be given soon in Sec. 2. The proof above is however the most satisfying one because it reduces the problem to an old legend: if a coin is tossed indefinitely, sooner or later it will be heads!

Exercise 1. What can one conclude by letting $a \to -\infty$ in (1.6) while keeping x and b fixed? Figure out a simple Markov process (not Brownian motion) for which (1.6) is true but (1.11) is false.

Exercise 2. Show that (1.11) remains true for $x = b$, a moot point in the proof given above.

2. Time and Place

The results in Sec. 1 are fundamental but qualitative. We now proceed to obtain quantitative information on the time τ and the place $x(\tau)$ of the exit from (a, b). Let us define

$$p_a(x) = P^x\{X(\tau) = a\} = P^x\{T_a < T_b\};$$
$$p_b(x) = P^x\{X(\tau) = b\} = P^x\{T_b < T_a\}. \tag{2.1}$$

It is a consequence of (1.7) that

$$p_a(x) + p_b(x) = 1, \qquad \forall x \in (a, b). \tag{2.2}$$

In order to solve for $p_a(x)$ and $p_b(x)$ we need another equation. The neatest way is to observe that the Brownian motion process is a martingale. More pedantically, let F_t be the σ-field generated by $\{x_s, 0 \leq s \leq t\}$, then for each $x \in R$:

$$\{X_t, F_t, P^x\} \text{ is a martingale.} \tag{2.3}$$

We leave the verification to the reader. The fundamental martingale stopping theorem then asserts that

$$\{X(t \wedge \tau), F(t \wedge \tau), P^x\} \qquad \text{is a martingale.} \tag{2.4}$$

The defining property of a martingale now yields

$$E^x\{X(0)\} = E^x\{X(t \wedge \tau)\}. \tag{2.5}$$

The left member above is equal to x; the right member is an "incalculable" quantity. Fortunately we can easily calculate its limit as $t \to \infty$. For almost every sample point ω, $\tau(\omega) < \infty$ by (1.6) and $t \wedge \tau(\omega) = \tau(\omega)$ for $t \geq \tau(\omega)$, hence $\lim_{t \to \infty} X(t \wedge \tau) = X(\tau)$ without even the continuity of $X(\cdot)$. Since

$$\sup_{0 \leq t < \infty} |X(t \wedge \tau)| \leq |a| \vee |b|, \tag{2.6}$$

the dominated convergence theorem allows us to take the limit as $t \to \infty$ under the E^x in (2.5) to conclude

$$x = E^x\{X(\tau)\}. \tag{2.7}$$

The novice must be warned that the verification of domination is absolutely *de rigueur* in such limit–taking, neglect of which has littered the field with published and unpublished garbage. [On this particular occasion the domination is of course trivial, but what if τ is replaced by T_a for instance?]

Since $X(\tau)$ takes only the two values a and b, (2.7) becomes

$$x = ap_a(x) + bp_b(x).$$ (2.8)

Solving (2.2) and (2.8) we obtain

$$p_a(x) = \frac{b-x}{b-a}, \quad p_b(x) = \frac{x-a}{b-a}, \quad x \in (a,b).$$ (2.9)

Note that (2.9) is valid only for $x \in [a,b]$, and implies $p_a(a) = 1$.

To obtain $E^x\{\tau\}$ we use another important martingale associated with the process:

$$\{X(t)^2 - t, F_t, P^x\} \text{ is a martingale}.$$ (2.10)

Application of the stopping theorem gives

$$x^2 = E^x\{X(\tau \wedge t)^2 - (\tau \wedge t)\}.$$

Since $X(\tau \wedge t)^2 \leq a^2 + b^2$ and $E^x\{\tau\} < \infty$ (see Sec. 1) we can let $t \to \infty$ and use dominated convergence (how?) to deduce

$$x^2 = E^x\{X(\tau)^2 - \tau\} = a^2 p_a(x) + b^2 p_b(x) - E^x\{\tau\}.$$ (2.11)

Together with (2.9) this yields

$$E^x\{\tau_{(a,b)}\} = (x-a)(b-x),$$ (2.12)

for all $x \in [a,b]$. For $x = a$ we have $E^a\{\tau_{(a,b)}\} = 0$, so $P^a\{\tau_{(a,b)} = 0\} = 1$.

From the last result it is tempting but fallacious to conclude that $P^a\{T_a = 0\} = 1$. In the next exercise we give a correct proof of this fact using the symmetry of Brownian motion, leaving the details to the reader.

Exercise 3. Prove the fundamental result

$$P^x\{T_x = 0\} = 1, \qquad \text{for all } x \in R. \qquad (2.13)$$

We may take $x = 0$. Define $T^- = \inf\{t > 0 : X(t) < 0\}$ and $T^+ = \inf\{t > 0 : X(t) > 0\}$. From the results above deduce $P^0\{T^- = 0\} = 1$ and $P^0\{T^+ = 0\} = 1$; hence $P^0\{T_0 = 0\} = 1$.

Exercise 4. Show that starting at any x, the Brownian path immediately crosses the x-level infinitely many times.

Exercise 5. Show that for any $x \neq b$ we have $E^x\{T_b\} = +\infty$.

Exercise 6. Let $x_n \to x$, then $P^x\{T_{x_n} \to 0\} = 1$.

Exercise 7. Let $\tau = \tau_{(a,b)}$, $t > 0$ and $\varphi(x) = P^x\{\tau > t\}$. Show that φ is concave in $[a, b]$ and hence continuous. Let x_1, x_2, $x \in [a, b]$ with $x = (1 - \lambda)x_1 + \lambda x_2$. Let $\tau' = \tau_{(x_1, x_2)}$. Check that $p_x(x_1) = 1 - \lambda$ and $p_x(x_2) = \lambda$, so it follows from the strong Markov property that

$$\varphi(y) \geq (1 - \lambda)\varphi(x_1) + \lambda\varphi(x_2).$$

Exercise 8. Use Exercise 6 and symmetry to show that for each t the maximum of $P^x\{\tau > t\}$ occurs at $x = (a + b)/2$.

3. A General Method

To obtain comprehensive information regarding the joint distribution of the time and place of exit from (a, b), we introduce a more powerful martingale.

Proposition 2. *For any real value of the parameter* α,

$$\left\{\exp\left(\alpha X(t) - \frac{\alpha^2 t}{2}\right), F_t, P^x\right\} \quad \text{is a martingale}. \qquad (3.1)$$

Proof. We begin with the formula

$$E^x\{\exp(\alpha x(t))\} = \exp\left(\alpha x + \frac{\alpha^2 t}{2}\right). \qquad (3.2)$$

The left member above being the probabilistic manifestation of the analytic formula

$$\int_{-\infty}^{\infty} e^{\alpha y} \frac{1}{\sqrt{2\pi t}} \exp\left\{-\frac{(x-y)^2}{2t}\right\} dy$$

its calculation is an exercise in beginner's calculus. When $\alpha = i\theta$ where $i = \sqrt{-1}$ and θ is real, (3.2) should be recognized as the familiar Fourier transform or characteristic function of the normal distribution commonly denoted by $N(x, t)$. If follows that if $0 \leq s < t$ then

$$E^x\left\{\exp\left(\alpha X(t) - \frac{\alpha^2 t}{2}\right)|F_s\right\}$$

$$= \exp\left(\alpha X(s) - \frac{\alpha^2 t}{2}\right) E^x\{\exp(\alpha(X(t) - X(s)))|F_s\}.$$

$$(3.3)$$

To be formal about it this time, let us put

$$\tilde{X}(u) = X(s + u) - X(s), \qquad \text{for } u \geq 0;$$

Since the Brownian motion is a process with stationary independent increments, the "shifted" process $\tilde{X}$ is a standard Brownian motion which is independent of F_s. Using (3.2) we have

$$E^x\{\exp(\alpha(X(t) - X(s)))|F_s\} = E^0\{\exp(\alpha X(t - s))\}$$

$$= \exp\left(\frac{\alpha^2(t - s)}{2}\right),$$

which confirms the assertion in (3.1).

For a few moments let us denote the martingale in (3.1) by $M(t)$. Then for every $x \in (a, b)$:

$$e^{\alpha x} = E^x\{M(0)\} = E^x\{M(t \wedge \tau)\}, \qquad (3.4)$$

where $\tau = \tau_{(a,b)}$. Since

$$|M(t \wedge \tau)| \le \exp(|\alpha|(|a| \wedge |b|)),$$

we obtain by dominated convergence just as in Sec. 2 that

$$e^{\alpha x} = E^x\{M(\tau)\} = E^x\left\{\exp\left(\alpha a - \frac{\alpha^2 \tau}{2}\right); X(\tau) = a\right\}$$

$$+ E^x\left\{\exp\left(\alpha b - \frac{\alpha^2 \tau}{2}\right); X(\tau) = b\right\}. \qquad (3.5)$$

Putting

$$f_a(x) = E^x\left\{\exp\left(-\frac{\alpha^2 \tau}{2}\right); X(\tau) = a\right\},$$

$$\qquad (3.6)$$

$$f_b(x) = E^x\left\{\exp\left(-\frac{\alpha^2 \tau}{2}\right); X(\tau) = b\right\},$$

we have the equation

$$e^{\alpha x} = e^{\alpha a} f_a(x) + e^{\alpha b} f_b(x), \quad x \in (a, b). \qquad (3.7)$$

We have also the equation

$$f_a(x) + f_b(x) = E^x\left\{\exp\left(-\frac{\alpha^2 \tau}{2}\right)\right\}. \qquad (3.8)$$

Unlike the situation in Sec. 2 these two equations do not yield the three unknowns involved. There are several ways of circumventing the difficulty. One is to uncover a third hidden equation — the reader should try to do so before peeking at the solution given below. But this quickie method depends on a lucky quirk. By contrast, the method developed here, though much longer, belongs to the mainstream of probabilistic analysis and is of wide applicability. It is especially charming in the setting of R^1.

We begin with the observation that if x is the midpoint of (a, b) then $f_a(x) = f_b(x)$ by symmetry so that in this case (3.7) is solvable for $f_a(x)$. Changing the notation we fix x and consider the interval $(x - h, x + h)$. We obtain from (3.7)

$$f_{x-h}(x) = \frac{e^{\alpha x}}{e^{\alpha(x-h)} + e^{\alpha(x+h)}} = \frac{1}{e^{\alpha h} + e^{-\alpha h}}$$

and consequently by (3.8)

$$E^x \left\{ \exp\left(-\frac{\alpha^2}{2} \tau_{(x-h, x+h)} \right) \right\} = \frac{1}{\operatorname{ch}(\alpha h)}. \qquad (3.9)$$

Here ch denotes the "hyperbolic cosine" function, as sh denotes the "hyperbolic sine" function:

$$\operatorname{ch} \theta = \frac{e^\theta + e^{-\theta}}{2}, \quad \operatorname{sh} \theta = \frac{e^\theta - e^{-\theta}}{2}. \qquad (3.10)$$

With this foot in the door, we will push on to calculate $f_a(x)$.

Recall $x \in (a, b)$, hence for sufficiently small $h > 0$ we have $[x - h, x + h] \subset (a, b)$ and so

$$\tau_{(x-h, x+h)} < \tau_{(a,b)}. \qquad (3.11)$$

We shall denote $\tau_{(x-h,x+h)}$ by $\tau(h)$ below; observe that it can also be defined as follows:

$$\tau(h) = \inf\{t > 0 : |X(t) - X(0)| \geq h\}, \qquad (3.12)$$

namely the first time that the path has moved a distance $\geq h$ (from whichever initial position). Now starting at x, the path upon its exit from $(x - h, x + h)$ will be at $x - h$ or $x + h$ with probability $1/2$ each. From the instant $\tau(h)$ onward, the path moves as if it started at these two new positions by the strong Markov property, because $\tau(h)$ is an optional time. This verbal description is made symbolic below.

$$E^x\left\{\exp\left(-\frac{\alpha^2\tau}{2}\right); X(\tau) = a\right\}$$

$$= E^x\left\{\exp\left(-\frac{\alpha^2\tau(h)}{2}\right); \qquad (3.13)\right.$$

$$\left. E^{X(\tau(h))}\left[\exp\left(-\frac{\alpha^2\tau}{2}\right); X(\tau) = a\right]\right\}.$$

It is crucial to understand why after the random shift of time given by $\tau(h)$, the "function to be integrated": $[\exp(-\frac{\alpha^2\tau}{2}); X(\tau) = a]$ does not change. This point is generally explained away by a tricky symbolism, but one should first perceive the truth with the naked eye. Anyway (3.13) may be written as

$f_a(x-h) = E^{(x-h)}\left(\exp\left(-\alpha^2\tau/2\right), X_{(\tau)} = a\right)$

$$f_a(x) = E^x\left\{\exp\left(-\frac{\alpha^2\tau(h)}{2}\right)\right\}\frac{1}{2}\{f_a(x-h) + f_a(x+h)\}.$$
$$(3.14)$$

Using (3.9) we may rewrite this as follows:

$$\frac{f_a(x+h) - 2f_a(x) + f_a(x-h)}{h^2} = \frac{2\mathrm{ch}(\alpha h) - 2}{h^2}f_a(x). \quad (3.15)$$

$\lim_{h \to 0}$ give $f''(x) = \alpha^2 f(x)$

$u'' - \alpha^2 u = 0$ has general sol'n

$u = A\, e^{-\alpha x} + B\, e^{\alpha x}.$

Letting $h \downarrow 0$ we see that the left member in (3.15) converges to $\alpha^2 f_a(x)$. It is also immediate from (3.14) that

$$f_a(x) < \frac{1}{2}\{f_a(x-h) + f_a(x+h)\}, \qquad (3.16)$$

valid for $a < x - h < x + h < b$. Since f_a is also bounded, (3.16) implies f_a is continuous in (a,b), in fact convex (see e.g.[Courant, Differential and Integral Calculus Vol. II, p. 326]). Now if f_a has a second derivative f_a'' in (a,b), then an easy exercise in calculus shows that the limit as $h \downarrow 0$ of the left member in (3.15) is equal to $f_a''(x)$. What is less easy is to show that a close converse is also true. This is known as Schwarz's theorem on generalized second derivative, a basic lemma in Fourier series. We state it in the form needed below.

Schwarz's Theorem. *Let f be continuous in (a,b) and suppose that*

$$\lim_{h \to 0} \frac{f(x+h) - 2f(x) + f(x-h)}{h^2} = \varphi(x) \quad \forall x \in (a,b)$$

where φ is continuous in (a,b). Then f is twice differentiable and $f'' = \varphi$ in (a,b).

It is sufficient to prove this theorem when $\varphi \equiv 0$ in (a,b), and a proof can be found in Titchmarsh, Theory of Functions, 2nd ed., p. 431. Since f_a has been shown to be continuous, Schwarz's theorem applied to (15) yields the differential equation:

$$f_a''(x) = \alpha^2 f_a(x), \qquad x \in (a,b). \qquad (3.17)$$

The most general solution of this equation is given by

$$f_a(x) = Ae^{\alpha x} + Be^{-\alpha x}, \qquad (3.18)$$

where A and B are two arbitrary constants. To determine these we compute the limits of $f_a(x)$ as $x \to a$ and $x \to b$, from inside (a, b). From (3.6) and (2.9) we infer that

$$\overline{\lim_{x \to b}} f_a(x) \le \lim_{x \to b} E^x\{X(\tau) = a\} = \lim_{x \to b} \frac{b - x}{b - a} = 0 \, ;$$

$$\overline{\lim_{x \to b}} f_b(x) \le \lim_{x \to b} E^x\{X(\tau) = b\} = \lim_{x \to b} \frac{x - a}{b - a} \le 1 \, . \qquad (3.19)$$

Since $f_a \ge 0$, the first relation above shows that $\lim_{x \to b} f_a(x) = 0$, using (3.7) we see that

$$e^{\alpha b} \le e^{\alpha b} \lim_{x \to b} f_b(x)$$

so $\lim_{x \to b} f_b(x) = 1$. Similarly we have

$$\lim_{x \to a} f_a(x) = 1 \, , \quad \lim_{x \to a} f_b(x) = 0 \, . \qquad (3.20)$$

Thus we obtain from (3.18):

$$0 = Ae^{\alpha b} + Be^{-\alpha b} \, , \quad 1 = Ae^{\alpha a} + Be^{-\alpha a} \, .$$

Solving for A and B and substituting into (3.18), we obtain

$$f_a(x) = \frac{\operatorname{sh} \alpha(b - x)}{\operatorname{sh} \alpha(b - a)} \, , \quad f_b(x) = \frac{\operatorname{sh} \alpha(x - a)}{\operatorname{sh} \alpha(b - a)} \, , \qquad (3.21)$$

where the second formula in (3.21) may be obtained from the first by interchanging a and b. Finally we have by (3.8):

$$E^x \left\{ \exp\left(-\frac{\alpha^2 \tau}{2} \right) \right\} = \frac{\operatorname{sh} \alpha(b - x) + \operatorname{sh} \alpha(x - a)}{\operatorname{sh} \alpha(b - a)} \, , \qquad (3.22)$$

Exercise 9. The quick way to obtain (3.21) is to use (3.7) for $-\alpha$ as well as $+\alpha$.

Exercise 10. Derive (2.9) from (3.21), and compute

$$E^x\{T_a; T_a < T_b\}.$$

Answer: $(b-x)(x-a)(2b-a-x)/3(b-a)$.

Exercise 11. Show that for $0 \le \theta < \pi^2/2(b-a)^2$, we have

$$E^x\{e^{\theta\tau(a,b)}\} = \frac{\cos\left(\sqrt{2\theta}\left(x - \frac{a+b}{2}\right)\right)}{\cos\left(\sqrt{2\theta}\left(\frac{b-a}{2}\right)\right)}.$$

Prove that $E^x\{e^{\theta\tau(a,b)}\} = +\infty$ for $\theta = \pi^2/2(b-a)^2$.

A third way to derive (3.21) will now be shown. Since (3.7) is valid for $-\alpha$ as well as α, letting $a \to -\infty$ we obtain

$$e^{|\alpha|x} = e^{|\alpha|b}E^x\left\{\exp\left(\frac{\alpha^2}{2}T_b\right)\right\}.$$

Changing $\alpha^2/2$ into λ, writing y for b and observing that the result is valid for any $x \neq y$ on account of symmetry of the Brownian motion with respect to right and left:

$$E^x\{e^{-\lambda T_y}\} = \exp(-\sqrt{2\lambda}|x-y|). \qquad (3.23)$$

This equation holds also when $x = y$, but the argument above does not include this case. The little sticking point returns to haunt us! We can dispose of it as follows. If $y \to x$ then $T_y \to T_x$ almost surely (proof?), hence the Laplace transform of T_y converges to that of T_x, and (3.23) takes on the limiting form

$$E^x\{e^{-\lambda T_x}\} = 1.$$

It follows that $P^x\{T_x = 0\} = 1$, namely Exercise 3 again.

Recalling that $\tau_{(a,b)} = T_a \wedge T_b$, we can write down the following relations:

$$E^x\{e^{-\lambda T_b}\}$$

$$= E^x\{\exp(-\lambda(T_a \wedge T_b)); T_b < T_a\}$$

$$+ E^x\{\exp(-\lambda\{T_a \wedge T_b\}); T_a < T_b\} \cdot E^a\{e^{-\lambda T_b}\}$$

$$E^x\{e^{-\lambda T_a}\}$$

$$= E^x\{\exp(\lambda(T_a \wedge T_b)); T_a < T_b\}$$

$$+ E^x\{\exp(-\lambda\{T_a \wedge T_b\}); T_b < T_a\} \cdot E^b\{e^{-\lambda T_a}\}.$$

(3.24)

Using (3.23) we see that these two equations can be solved for the two unknowns which are just $f_a(x)$ and $f_b(x)$ after a change of notation.

4. Drift

The methods above can be used to obtain analogous results for a Brownian motion with a constant *drift*, namely for the process:

$$\tilde{X}(t) = X(t) + ct \qquad (4.1)$$

where $X(t)$ is the standard Brownian motion and c is a nonzero constant. We may suppose $c > 0$ for definiteness.

The strong law of large numbers implies that almost surely

$$\lim_{t \to \infty} \tilde{X}(t) = +\infty. \qquad (4.2)$$

The argument in Sec. 1 is still valid to show that exit from any given interval (a, b) is almost sure, but the analogue to Proposition 1 must be false. The reader should find out for himself that the martingales in (2.3) and (2.10) translated in terms of $\tilde{X}$, are not sufficient to determine

$$\tilde{p}_a(x) = P^x\{\tilde{X}(\tilde{\tau}) = a\}, \quad \tilde{p}_b(x) = P^x\{\tilde{X}(\tilde{\tau}) = b\}, \qquad (4.3)$$

where

$$\tilde{\tau} = \tilde{\tau}_{(a,b)} = \inf \{t > 0 : \tilde{X}(t) \notin (a,b)\}.$$

Fortunately, the martingale in (3.1) can be manipulated to do so.

Take $a = -2c$ in (3.1). We have

$$\exp\left(-2cX(t) - \frac{(2c)^2 t}{2}\right) = \exp(-2c\tilde{X}(t)). \qquad (4.4)$$

Write $s(x) = e^{-2cx}$, then

$$\{s(\tilde{X}(t)), F_t, P^x\} \quad \text{is a martingale}. \qquad (4.5)$$

It follows that

$$s(x) = s(a)\tilde{p}_a(x) + s(b)\tilde{p}_b(x) \qquad (4.6)$$

together with

$$\tilde{p}_a(x) + \tilde{p}_b(x) = 1, \qquad (4.7)$$

we obtain

$$\tilde{p}_a(x) = \frac{s(b) - s(x)}{s(b) - s(a)}, \quad \tilde{p}_b(x) = \frac{s(x) - s(a)}{s(b) - s(a)}. \qquad (4.8)$$

The function s is called the scale function. Compare (1.8) with (1.9). We can now use the martingale $\tilde{X}(t) - ct$ to find $E^x\{\tilde{\tau}\}$. As before, the stopping theorem yields

$$x = E^x\{\tilde{X}(\tilde{\tau}) - c\tilde{\tau}\} = a\tilde{p}_a(x) + b\tilde{p}_b(x) - cE^x\{\tilde{\tau}\},$$

and so

$$E^x\{\tilde{\tau}\} = \frac{a(s(b) - s(x)) + b(s(x) - s(a)) - x(s(b) - s(a))}{c(s(b) - s(a))}.$$

More interesting is to use (4.8) to obtain information for the hitting time

$$T_y = \inf \{t > 0 : X(t) = y\}.$$

If we let $b \to \infty$ in the first and $a \to -\infty$ in the second equation in (4.8), the results are as follows:

$$P^x\{\tilde{T}_a < \infty\} = e^{-2c(x-a)}, \quad a < x < \infty;$$

$$P^x\{\tilde{T}_b < \infty\} = 1, \quad -\infty < x < b.$$

(4.9)

The second relation is of course an immediate consequence of (4.2), and the continuity of the paths.

To obtain the distribution of $\tilde{T}_y$, we return to the martingale in (3.1) translated in terms of $\tilde{X}(t)$;

$$\exp\left(a\tilde{X}(t) - \left(ac + \frac{a^2}{2}\right)t\right).$$

Put

$$\lambda = ac + \frac{a^2}{2},$$

$$\alpha = -c \pm \sqrt{2\lambda + c^2}.$$

we obtain in the usual manner

$$e^{\alpha x} = e^{\alpha a} E^x\{\exp(-\lambda\tilde{\tau}); \tilde{X}(\tilde{\tau}) = a\}$$

$$+ e^{\alpha b} E^x\{\exp(-\lambda\tilde{\tau}); \tilde{X}(\tilde{\tau}) = b\}.$$

(4.11)

Choose the $+$ sign in α so that $\alpha > 0$, and let $a \to -\infty$. Then choose the $-$ sign in α so that $\alpha < 0$, and let $b \to +\infty$. The results may be recorded as follows:

$$E^x\{\exp(-\lambda\tilde{T}_y)\} = \exp(-\sqrt{2\lambda + c^2}|x - y| - c(x - y)). \quad (4.12)$$

Using (3.24) we can obtain the joint distribution of $\tilde{X}(\tau)$ and $\tilde{\tau}$. In general the results are complicated but one interesting case emerges when $x = 0$, $b > 0$ and $a = -b$. In this case if we let $f_+(\lambda) = E^0(\exp(-\lambda\tilde{\tau}); \tilde{X}(\tilde{\tau}) = b)$ and $f_-(\lambda) = E^0(\exp(-\lambda\tilde{\tau}); \tilde{X}(\tilde{\tau}) = -b)$, then (3.24) becomes

$$\exp(-\sqrt{2\lambda + c^2}b + bc)$$
$$= f_+(\lambda) + f_-(\lambda)\exp(-\sqrt{2\lambda + c^2}(2b) + 2bc),$$
$$\exp(-\sqrt{2\lambda + c^2}b - bc)$$
$$= f_-(\lambda) + f_+(\lambda)\exp(-\sqrt{2\lambda + c^2}(2b) + 2bc).$$

(4.13)

Dividing each equation by its left hand side and subtracting, we obtain

$$f_+(\lambda)(\exp(\sqrt{2\lambda + c^2}b - bc) - \exp(-\sqrt{2\lambda + c^2}b - bc))$$
$$= f_-(\lambda)(\exp(\sqrt{2\lambda - c^2}b + bc) - \exp(-\sqrt{2\lambda + c^2}b + bc)),$$

and consequently

$$f_+(\lambda) = e^{2bc}f_-(\lambda).$$

(4.14)

Since we have also from (4.8)

$$P^0(\tilde{X}(\tilde{\tau}) = b) = e^{2bc}P^0(\tilde{X}(\tilde{\tau}) = -b),$$

(4.15)

it follows that

$$E^0(\exp(-\lambda\tilde{\tau})|\tilde{X}(\tilde{\tau}) = b) = E^0(\exp(\lambda\tilde{\tau})|\tilde{X}(\tilde{\tau}) = -b).$$

That is, the exit time $\tilde{\tau}$ and the exit place $\tilde{X}(\tilde{\tau})$ are independent. This curious fact was first observed by Frederick Stein.* Is there an intuitive explanation?

Exercise 12. Almost every sample function $\tilde{X}(\cdot, \omega)$ has a minimum value $m(\omega) > -\infty$. Use the strong Markov property to show that m has an exponential distribution, and then find this distribution.

Exercise 13. Show that almost every path of $\tilde{X}$ reaches its minimum value $m(\omega)$ only once.

Exercise 14. This exercise, which is based on a result of J. W. Pitman and J. C. Rogers, shows that sometimes processes which are "obviously" not Markovian actually are. Let X^+ and X^- be independent Brownian motions with drifts $+c$ and $-c$ respectively and let ξ be an independent random variable which $= +1$ with probability p and $= -1$ with probability $1 - p$. Construct a process Y by letting $Y_t = X_t^+$ on $\{\xi = 1\}$ and $Y_t = X_t^-$ on $\{\xi = -1\}$. The claim is that Y is a Markov process with respect to G_t, the σ-field generated by Y_t, $s \leq t$.

At first glance this seems false because watching Y_t gives us information about ξ which can be used to predict the future development of the process. This is true but a little more thought shows

$$P^0(\xi = 1 | G_t) = e^{cx_t} / (e^{cX_t} + e^{-cX_t}).$$

Verify this by (4.9) and show that Y_t is Markovian. [I owe this exercise to R. Durrett.]

---

*"An independence in Brownian motion with constant drift", *Ann. of Prob.* 5 (1977), 571–572.

5. Dirichlet and Poisson Problems

In classical potential theory (see Kellogg [1]) there are a clutch of famous problems which had their origins in electromagnetism. We begin by stating two of these problems in Euclidean space R^d, where d is the dimension. Let D be a nonempty bounded open set (called a "domain" when it is connected), and let ∂D denote its boundary: $\partial D = \bar{D} \cap \overline{(D^c)}$ where the upper bar denotes closure. Let Δ denote the Laplacian, namely the differential operator

$$\Delta = \sum_{j=1}^{d} \left(\frac{\partial}{\partial x_j} \right)^2 . \tag{5.1}$$

A function defined in D is called *harmonic* there iff $\Delta f = 0$ in D. This of course requires that f is twice differentiable. If f is locally integrable in D, namely has a finite Lebesgue integral over any compact subset of D, then it is harmonic in D if and only if the following "surface averaging property" is true. Let $B(x, \delta)$ denote the closed ball with center x and radius δ. For each $x \in D$ and $\delta > 0$ such that $B(x, \delta) \subset D$, we have

$$f(x) = \frac{1}{\sigma(\partial B(x, \delta))} \int_{\partial B(x,\delta)} f(y)\sigma(dy) \tag{5.2}$$

where $\sigma(dy)$ is the area measure on $\partial B(x, \delta)$. This alternative characterization of harmonic function is known as Gauss's theorem and plays a basic role in probabilistic potential theory, because probability reasoning integrates better than differentiates.

Dirichlet's problem (or first boundary value problem). Given D and a continuous function f on ∂D, to find a function φ which is continuous in $\bar{D}$ and satisfies:

$$\Delta \varphi = 0 \quad \text{in} \quad D,$$
$$\varphi = f \quad \text{on} \quad \partial D. \tag{5.3}$$

Poisson's problem. Given D and a continuous function f in D, to find a function φ which is continuous in D and satisfies

$$\Delta \varphi = f \quad \text{in} \quad D,$$
$$\varphi = 0 \quad \text{on} \quad \partial D. \tag{5.4}$$

We have stated these problems in the original forms, of which there are well-known generalizations. As stated, a unique solution to either problem exists provided that the boundary ∂D is not too irregular. Since we shall treat only the one-dimensional case we need not be concerned with the general difficulties.

In R^1, a domain is just an bounded open nonempty interval $I = (a, b)$. Its boundary ∂I consists of the two points $\{a, b\}$. Since $\Delta f = f''$, a harmonic function is just a linear function. The boundary function f reduces to two arbitrary values assigned to the points a and b, and no question of its continuity arises. Thus in R^1 Dirichlet's problem reads as follows.

Problem 1. Given two arbitrary numbers $f(a)$ and $f(b)$, to find a function φ which is linear in (a, b) and continuous in $[a, b]$, such that $\varphi(a) = f(a)$, $\varphi(b) = f(b)$.

This is a (junior) high school problem of analytic geometry. The solution is given by

$$\frac{b-x}{b-a} f(a) + \frac{x-a}{b-a} f(b). \tag{5.5}$$

Now we will write down the probabilistic solution, as follows

$$\varphi(x) = E^x\{f(X(\tau))\}, \quad x \in (a, b) \tag{5.6}$$

where $\tau = \tau_{(a,b)}$. If we evaluate the right member of (5.6) by (2.9), we see at once that it is the same as given in (5.5). But we will prove that φ is the sought solution by the general method developed in Sec. 3, because the same pattern of proof works in any dimension. Using the $\tau(h)$ of (3.12), we obtain

$$\varphi(x) = E^x\{E^{X(\tau(h))}[f(X(\tau))]\} = \frac{1}{2}\{\varphi(x - h) + \varphi(x + h)\} \tag{5.7}$$

for any h for which (3.11) is true. This is the one-dimensional case of Gauss's criterion for harmonicity. Since φ is bounded it follows from the criterion that φ is harmonic, namely linear. But we can also invoke Schwarz's Theorem in Sec. 3 to deduce this result, indeed the generalized second derivative of φ is identically zero by (5.7)

It remains to show that as $x \to a$ or b from inside (a, b), $\varphi(x)$ tends to $f(a)$ or $f(b)$ respectively. This is a consequence of the probabilistic relations below:

$$\lim_{x \to a} P^x\{\tau = T_a\} = 1, \quad \lim_{x \to b} P^x\{\tau = T_b\} = 1 \tag{5.8}$$

which are immediate by (2.9). But since no such analogue is available in dimension > 1, another proof more in the general spirit is indicated in Exercise 15 below. Assuming (5.8), we have

$$\varphi(x) = E^x\{f(X(T_a)); \tau = T_a\} + E^x\{f(X(T_b)); \tau = T_b\}$$

$$= P^x\{\tau = T_a\}f(a) + P^x\{\tau = T_b\}f(b),$$

and consequently

$$\lim_{x \to a} \varphi(x) = 1 \cdot f(a) + 0 \cdot f(b) = f(a);$$

$$\lim_{x \to b} \varphi(x) = 0 \cdot f(a) + 1 \cdot f(b) = f(b).$$

Thus the extension of φ to $[a, b]$ agrees with f at a and b. [Since φ is linear in (a, b), it has a trivial continuous extension to $[a, b]$. This no longer trivial in dimension > 1.]

Exercise 15. Show that for any $\varepsilon > 0$,

$$\lim_{x \to 0} P^x \{T_0 \le \varepsilon\} = 1.$$

This is equivalent to $\lim_{x \to 0} P^0 \{T_{-x} \le \varepsilon\} = 1$, and is a case of Exercise 6. Now derive (5.8) from (5.9).

Problem 2. Given a bounded continuous function f in (a, b), to find a function φ which is continuous in $[a, b]$ such that

$$\frac{1}{2}\varphi''(x) = -f(x), \quad \text{for } x \in (a, b),$$

$$\varphi(a) = \varphi(b) = 0.$$

(5.10)

The constants $\frac{1}{2}$ and -1 in the differential equation are chosen for the sake of convenience, as will become apparent below. This is a simple calculus problem which can be solved by setting

$$\varphi(x) = \int_a^x 2(y - x)f(y)dy + cx + d$$

and determining the constants $d = -ca$ and

$$c = (b-a)^{-1} \int_a^b (b-y)f(y)dy$$

by the boundary conditions $\varphi(a) = 0$ and $\varphi(b) = 0$. Substituting these values for c and d and rearranging we can write the solution above as

$$\varphi(x) = \int_a^b g(x,y)f(y)dy \qquad (5.11)$$

where

$$g(x,y) = \begin{cases} \dfrac{2(x-a)(b-y)}{b-a}, & \text{if } a < x \le y < b, \\[2mm] \dfrac{2(b-x)(y-a)}{b-a}, & \text{if } a < y \le x < b. \end{cases} \qquad (5.12)$$

Note that $g(x,y) > 0$ in (a,b) and $g(x,y) = g(y,x)$. We put $g(x,y) = 0$ outside $(a,b) \times (a,b)$. The function g is known as the Green's function for (a,b) because representing the solution of (5.10) in the form (5.11) is an example of the classical method of solving differential equations by Green's functions (see Courant and Hilbert [2] Ch. V. 14 and Exercises 17 and 18 below).

Now we will write down the probabilistic solution of Problem 2, as follows:

$$\varphi(x) = E^x \left\{ \int_0^\tau f(X(t))dt \right\}, \qquad (5.13)$$

Note that the integral above may be regarded as over $(0,\tau)$ so that f need be defined in (a,b) only. Without loss of generality we may suppose $f \geq 0$; for the general case will follow from this case and $f = f^+ - f^-$. To show that φ satisfies the differential equation, we proceed by the method of Sec. 3, we have

$$\varphi(x) = E^x \left\{ \left(\int_0^{\tau(h)} + \int_{\tau(h)}^{\tau} \right) f(X(t)) dt \right\}$$

$$= E^x \left\{ \int_0^{\tau(h)} f(X(t)) dt \right\}$$

$$+ E^x \left\{ E^{X(\tau(h))} \left[\int_0^{\tau} f(X(t)) dt \right] \right\}. \qquad (5.14)$$

Let us put

$$\psi(x, h) = E^x \left\{ \int_0^{\tau(h)} f(X(t)) dt \right\}, \qquad (5.15)$$

then

$$\varphi(x) = \psi(x, h) + \frac{1}{2} \{ \varphi(x + h) + \varphi(x - h) \}. \qquad (5.16)$$

Since $f \geq 0$, $\psi \geq 0$; also $\varphi(x) \leq ||f|| E^x \{\tau\} \leq ||f|| (b - a)^2 / 4$. Thus φ is continuous and concave. Now write (5.16) as

$$\frac{\varphi(x + h) - 2\varphi(x) + \varphi(x - h)}{h^2} = -\frac{2\psi(x, h)}{h^2}. \qquad (5.17)$$

To calculate the limit of the right member of (5.17) as $h \to 0$, we note by (2.12):

$$E^x \{\tau(h)\} = h^2. \qquad (5.18)$$

Next we have

$$\psi(x,h) - f(x)E^x\{\tau(h)\} = E^x\left\{\int_0^{\tau(h)} [f(X(t)) - f(X(0))]dt\right\}.$$
$$(5.19)$$

Since f is continuous at x, given $\varepsilon > 0$ there exists $h_0(\varepsilon)$ such that if $|y - x| \le h_0(\varepsilon)$ then $|f(y) - f(x)| \le \varepsilon$. Hence if $0 < h < h_0$, we have $|f(X(t)) - f(X(0))| \le \varepsilon$ for $0 \le t \le \tau(h)$ and so the absolute value of the right member of (5.19) is bounded by $E^x\{\varepsilon\tau(h)\} = \varepsilon h^2$. It follows that the left member of (5.19) divided by h^2 converges to zero as $h \to 0$, and consequently by (5.18)

$$\lim_{h \to 0} \frac{\psi(x,h)}{h^2} = f(x). \qquad (5.20)$$

Since φ is continuous by concavity from (5.16), and f is continuous by hypothesis, an application of Schwarz's Theorem yields the desired result

$$\varphi''(x) = -2f(x).$$

Furthermore since

$$|\varphi(x)| \le ||f||E^x\{\tau\},$$

$\varphi(x)$ converges to zero as $x \to a$ or $x \to b$ by (2.12). On the other hand $\varphi(a) = \varphi(b) = 0$ by (2.13). Thus φ is continuous in $[a, b]$ and vanishes at the endpoints.

If we equate the two solutions of Problem 2 given in (5.12) and (5.13), we obtain

$$E^x\left\{\int_0^\tau f(X(t))dt\right\} = \int_a^b g(x,y)f(y)dy \qquad (5.21)$$

for every bounded continuous f on (a, b). Let us put for $x \in R^1$ and $B \in \mathcal{B}^1$:

$$V(x, B) = E^x \left\{ \int_0^\tau 1_B(X(t)) dt \right\}. \qquad (5.22)$$

Then it follows from (5.21) and F. Riesz's theorem on the representation of linear functionals on (a, b) as measures (see, e.g., Royden [4, p. 310]) that we have

$$V(x, B) = \int_B g(x, y) dy. \qquad (5.23)$$

In other words, $V(x, \cdot)$ has $g(x, \cdot)$ as its Radon–Nikodym derivative with respect to the Lebesgue measure on (a, b). The kernel V is sometimes called the potential of the Brownian motion killed at τ. It is an important object for the study of this process since $V(x, B)$ gives the expected occupation time of B starting form x.

Exercise 16. Show by using elementary calculus that the solutions to Problems 1 and 2 in R^1 are unique.

Exercise 17. Define a function $g(x, y)$ in $[a, b]$ as follows. For each x let $g_x(\cdot) = g(x, \cdot)$.

  (i) g_x is a continuous function with $g_x(a) = g_x(b) = 0$;
  (ii) for all $x \neq y$, $g_x''(y) = 0$,
  (iii) $\lim_{\varepsilon \to 0} (g_x'(x + \varepsilon) - g_x'(x - \varepsilon)) = -1$.

Show that the function $g(x, y)$ defined by (5.12) is the only function with these properties.

Example 18. Let the function δ_y have the defining property that for any function f on (a, b) we have

$$\int_a^b \delta_y(u) f(u) du = f(y). \tag{5.23}$$

This δ_y is called the **Dirac delta function** [never mind its existence!]. It follows that

$$\int_a^x \delta_y(u) du = 1_{(a,x)}(y). \tag{5.24}$$

We can now solve the differential equation

$$h'' = -2\delta_y, \quad h(a) = h(b) = 0.$$

by another integration of (5.24). Carry this out to obtain $h(x) = g(x, y)$, which is what results if we let $f = -2\delta_y$ in (5.21).

Exercise 19. Determine the measure $H(x, \cdot)$ on ∂I so that the solution to Problem 1 may be written as

$$\int_{\partial I} f(y) H(x, dy).$$

The analogue in R^d is called the harmonic measure for I. It is known in the classical theory that this measure may be obtained by taking the "interior normal derivative" of $g(x, y)$ with respect to y. Find out what this means in R^1.

Exercise 20. Give meaning to the inverse relations:

$$\frac{\Delta}{2}(-G) = I, \quad (-G)\frac{\Delta}{2} = I$$

where I is the identity, and G is the operator defined by $Gf(x) = \int_a^b g(x,y)f(y)dy$.

Exercise 21. Solve the following problem which is a combination of Problems 1 and 2. Given f_2 on ∂I and continuous f_1 in I, find φ such that φ is continuous in I and satisfies

$$\frac{1}{2}\varphi'' = -f_1 \quad \text{in } I,$$

$$\varphi = f_2 \quad \text{on } \partial I.$$

6. Feynman–Kac Functional

As a final application of the general method, we will treat a fairly new problem. Reversing the previous order of discussion, let us consider

$$\varphi(x) = E^x \left\{ \exp \int_0^\tau q(X(t))dt \cdot f(X(\tau)) \right\}, \quad x \in [a,b] \quad (6.1)$$

where q is a bounded continuous function in $[a,b]$, f as in Problem 1 (6.1). Note that by Exercise 3:

$$\varphi(a) = f(a), \qquad \varphi(b) = f(b).$$

The exponential factor in (6.1) is called the Feynman–Kac functional.

An immediate question is whether φ is finite. If $q \equiv a$ constant c, and $f \equiv 1$, then $\varphi \equiv \infty$ for sufficiently large c, by Exercise 11.

Let us write $e(u) = \int_0^u q(X(t))dt$ for $u \geq 0$.

Proposition 3. *Suppose $f \geq 0$ in (6.1). If $\varphi \not\equiv \infty$ in (a,b), then φ is continuous in $[a,b]$.*

Proof. Let $\varphi(x_0) < \infty$, and $x \neq x_0$, $x \in (a, b)$. Then we have by the strong Markov property

$$\infty > \varphi(x_0) \geq E^{x_0}\{e(\tau); T_x < \tau\} = E^{x_0}\{e(T_x); T_x < \tau\}\varphi(x).$$

Since $P^{x_0}\{T_x < \tau\} > 0$ and $e(T_x) > 0$, this implies $\varphi(x) < \infty$.

Next, given any $A > 0$, there exists $\delta > 0$ such that we have

$$E^x\{e^{A\tau(\delta)}\} < \infty. \tag{6.3}$$

This follows from the derivation of (1.7). Consequently we have by dominated convergence

$$\lim_{h \to 0} E^x\{e^{A\tau(h)}\} = 1. \tag{6.4}$$

We now state a lemma.

Lemma. *Let φ be a finite nonnegative function on $[a, b]$ having the following approximate convexity property. For each $[x_1, x_2] \subset [a, b]$, $0 < x_2 - x_1 < \delta(\varepsilon)$ and $x = \lambda x_1 + (1 - \lambda)x_2$, $0 < \lambda < 1$, then*

$$(1 - \varepsilon)\{\lambda\varphi(x_1) + (1 - \lambda)\varphi(x_2)\}$$

$$< \varphi(x) < (1 + \varepsilon)\{\lambda\varphi(x_1) + (1 - \lambda)\varphi(x_2)\}. \tag{6.5}$$

Such a φ is continuous in $[a, b]$.

Proof of the lemma. Let $\varepsilon = 1$, $h < \delta(1)$ and $\lambda = \frac{1}{2}$, then we have by (6.5)

$$\varphi(x) < \{\varphi(x + h) + \varphi(x - h)\}. \tag{6.6}$$

For a suitable h we can divide $[a, b]$ into a finite number of subintervals of length $2h$ each. If we apply (6.5) to each subinterval we see that ϕ is bounded.

Now fix x in (a, b) and shrink $[x_1, x_2]$ to x in such a way that $\lambda \to 1$ and $\varphi(x_1) \to \lim_{y \uparrow \uparrow x} \varphi(y)$ or $\varphi(x_1) \to \overline{\lim}_{y \uparrow \uparrow x} \varphi(y)$, we obtain from (6.5):

$$(1 - \varepsilon) \overline{\lim_{y \uparrow \uparrow x}} \varphi(y) < \varphi(x) < (1 + \varepsilon) \lim_{y \uparrow \uparrow x} \varphi(y) .$$

Similarly for $y \downarrow \downarrow x$. Since ε is arbitrary this shows that φ is continuous at x. A similar argument shows that φ is unilaterally continuous at a and at b. Lemma is proved.

We return to φ and generalize the basic argument in Sec. 3 by considering the first exit time from an asymmetric interval $(x - h, x + h') \subset (a, b)$, starting from x. Recall that

$$P^x \{T_{x-h} < T_{x+h'}\} = \frac{h'}{h + h'} . \tag{6.7}$$

For sufficiently small h and h' we have by (6.4):

$$1 - \varepsilon < E^x \{e^{-Q\tau^*}\}, \quad E^x \{e^{Q\tau^*}\} < 1 + \varepsilon \tag{6.8}$$

where $Q = \|q\|$ and $\tau^* = \tau_{(x-h \ x+h')}$. The strong Markov property yields, for the φ in (6.1) with an arbitrary f:

$$\varphi(x) = E^x \{e(\tau^*)\} \left\{ \frac{h'}{h + h'} \varphi(x - h) + \frac{h}{h + h'} \varphi(x + h') \right\} . \tag{6.9}$$

Using (6.8) and (6.9) we see that if $f \geq 0$ in (6.1) then φ satisfies the conditions of the Lemma, and is therefore continuous in $[a, b]$. In particular this is true when $f = 1_{\{a\}}$ or $1_{\{b\}}$. Hence it is also true for the φ in (6.1) for an arbitrary finite f. Proposition 3 is proved.

Let us write φ_a and φ_b for the φ in (6.1) when $f = 1_{\{a\}}$ and $f = 1_{\{b\}}$ respectively. According to Proposition 3, either

$\varphi_a \equiv \infty$ or φ_a is bounded continuous in $[a, b]$, and similarly for φ_b. However, it seems possible that $\varphi_a \equiv \infty$ but $\varphi_b \not\equiv \infty$ in $[a, b]$, or vice versa. For a general f, we have

$$\varphi(x) = f(a)\varphi_a(x) + f(b)\varphi_b(x), \tag{6.10}$$

provided the right member above is not $+\infty - \infty$ or $-\infty + \infty$. This is certainly the case under the hypothesis of the next proposition.

Proposition 4 *Suppose that $\varphi_a \not\equiv \infty$ and $\varphi_b \not\equiv \infty$ in (a, b). Then for any $f \geq 0$ we have*

$$\frac{1}{2}\varphi'' + q\varphi = 0$$

in (a, b), and φ is continuous in $[a, b]$.

Proof. Write

$$E^x\{e(\tau(h))\} = 1 + \psi(x, h); \tag{6.11}$$

then (6.9) for $h = h'$ takes the form:

$$\frac{\varphi(x+h) - 2\varphi(x) + \varphi(x-h)}{h^2} = -\frac{\psi(x, h)}{h^2}\{\varphi(x+h) + \varphi(x-h)\}. \tag{6.12}$$

Since we have proved that φ is continuous, the quantity in (6.12) will converge to $-\lim_{h \to 0}[\psi(x, h)/h^2]2\varphi(x)$ as $h \to 0$, provided that the latter limit exists. To show this we need

$$E^x\{\tau(h)^2\} = \frac{5}{3}h^4; \tag{6.13}$$

also that for sufficiently small h we have by (6.4)

$$E^x\{e^{4(Q+1)\tau(h)}\} \leq 2. \tag{6.14}$$

Exercise 22. Prove (6.13). Can you get a general formula for $E^x\{\tau(h)^k\}$, $k \geq 1$?

Using the trivial inequality $\sqrt{u} \leq e^u$ for all $0 \leq u < \infty$, we have

$$\tau(h)^2 e^{Q\tau(h)} \leq \tau(h)^{3/2} e^{(Q+1)\tau(h)} .$$

Hence by Hölder's inequality (6.13) and (6.14)

$$E^x\{\tau(h)^2 e^{Q\tau(h)}\} \leq E^x\{\tau(h)^2\}^{3/4} E^x\{e^{4(Q+1)\tau(h)}\}^{1/4} \leq c_1 h^3 , \tag{6.15}$$

where c_1 is a constant. Next we use the inequality

$$|e^u - 1 - u| \leq \frac{u^2}{2} e^{|u|} ,$$

valid for all u, to obtain

$$E^x\left\{\left|e(\tau(h)) - 1 - \int_0^{\tau(h)} q(X(t))dt\right|\right\}$$

$$\leq \frac{1}{2} E^x\left\{\left(\int_0^{\tau(h)} q(X(t))dt\right)^2 e^{Q\tau(h)}\right\}$$

$$\leq \frac{Q^2}{2} E^x\{\tau(h)^2 e^{Q\tau(h)}\} .$$

The last term divided by h^2 converges to zero as $h \to 0$, by (6.15). Hence by (5.20) with f replaced by q:

$$\lim_{h \to 0} \frac{\psi(x, h)}{h^2} = \lim_{h \to 0} \frac{1}{h^2} E^x\left\{\int_0^{\tau(h)} q(X(t))dt\right\} = q(x) .$$

Therefore Schwarz's theorem applied to (6.12) yields $\varphi'' = -2q$ as asserted. Note that the continuity of φ is required here also. Proposition 4 is proved.

Propositions 3 and 4 together give a dichotomic criterion for the solvability of the following problem.

Problem 3. Given a bounded continuous function q in (a, b) and two arbitrary number $f(a)$ and $f(b)$, to find a function φ which is continuous in $[a, b]$ such that

$$\frac{1}{2}\varphi''(x) + q(x)\varphi(x) = 0, \quad x \in (a, b);$$
$$\varphi(a) = f(a), \quad \varphi(b) = f(b). \tag{6.16}$$

Exercise 23. Is the solution to Problem 3 unique when it exists?

Exercise 24. Solve the problem similar to Problem 3 but with the right side of the differential equation in (6.16) replaced by a given bounded continuous function in (a, b). This is the Poisson problem with the Feynman–Kac functional.

Exercise 25. Prove that if the equation in (6.16) has a positive solution φ in (a, b), then for any $[c, d] \subset (a, b)$, we have

$$\varphi(x) = E^x\{e(\tau_{(c,d)})\varphi(X(\tau_{(c,d)}))\}, \quad x \in [c, d].$$

In particular,

$$x \to E^x\{e(\tau_{c,d})\}$$

is bounded in $[c, d]$.

Exercise 26. Prove that if the differential equation in (6.16) has a positive solution in each interval (c, d) such that $[c, d] \subset (a, b)$ (without any condition on the boundary $\{c, d\}$) then it has a positive solution in (a, b). These solutions are a prior unrelated to one another.

Exercise 27. Is it possible that $\varphi_a \equiv \infty$ in (a, b) whereas $\varphi_b \not\equiv \infty$ in (a, b)? Here φ_a and φ_b are defined before Proposition 4. This is a very interesting problem solved by M. Hogan, a graduate student in my class.

REFERENCES

[1] Kellogg, see under References in Part I.

[2] Courant and Hilbert, ditto.

[3] Chung, ditto.

[4] Royden, H., Real Analysis (second edition) Macmillan, New York, 1968.

Part III.

Stopped Feynman–Kac Functionals

1. Introduction*

Let $X = \{X(t), t \geq 0\}$ be a strong Markov process with continuous paths on $R = (-\infty, +\infty)$. Such a process is often called a diffusion. For each real b, we define the hitting time τ_b as follows:

$$\tau_b = \inf\{t > 0 \mid X(t) = b\}.$$

Let P_a and E_a denote as usual the basic probability and expectation associated with paths starting from a. It is assumed that for every a and b, we have

$$P_a\{\tau_b < \infty\} = 1. \tag{1}$$

Now let q be a bounded Borel measurable function on R, and write for brevity

$$e(t) = \exp\left(\int_0^t q(X(s))ds\right). \tag{2}$$

* Séminaire de probabilités, XIV (1978/9), Strasbourg, France.

This is a multiplicative functional introduced by R. Feynman and M. Kac. In this paper we study the quantity

$$u(a,b) = E_a\{e(\tau_b)\}. \tag{4}$$

Since q is bounded below, (2) implies that $u(a,b) > 0$ for every a and b, but it may be equal to $+\infty$. A fundamental property of u is given by

$$u(a,b)u(b,c) = u(a,c), \tag{5}$$

valid for $a < b < c$, or $a > b > c$. This is a consequence of the strong Markov property (SMP).

2. The Results

We begin by defining two abscissas of finiteness, one for each direction.

$$
\begin{aligned}
\beta &= \inf\{b \in R \,|\, \exists a < b : u(a,b) = \infty\} \\
&= \sup\{b \in R \,|\, \forall a < b : u(a,b) < \infty\}; \\
\alpha &= \sup\{a \in R \,|\, \exists b > a : u(b,a) = \infty\} \\
&= \inf\{a \in R \,|\, \forall b > a : u(b,a) < \infty\}.
\end{aligned} \tag{6}
$$

It is possible, e.g., that $\beta = -\infty$ or $+\infty$. The first case occurs when X is the standard Brownian motion, and $q(x) \equiv 1$; for then, $u(a,b) \geq E_a(\tau_b) = \infty$, for any $a \neq b$.

Lemma 1. *We have*

$$
\begin{aligned}
\beta &= \inf\{b \in R \,|\, \forall a < b : u(a,b) = \infty\} \\
&= \sup\{b \in R \,|\, \exists a < b : u(a,b) < \infty\}; \\
\alpha &= \sup\{a \in R \,|\, \forall b > a : u(b,a) = \infty\} \\
&= \inf\{a \in R \,|\, \exists b > a : u(b,a) < \infty\}.
\end{aligned}
$$

Proof. It is sufficient to prove the first equation above for β, because the second is trivially equivalent to it, and the equations for α follow by similar arguments. Suppose $u(a,b) = \infty$; then for $x < a < b$ we have $u(x,b) = \infty$ by (5). For $a < x < b$ we have by SMP,

$$u(x,b) \geq E_x\{e(\tau_a)\,;\tau_a < \tau_b\}u(a,b) = \infty$$

since $P_x\{\tau_a < \tau_b\} > 0$ in consequence of (2).

The next lemma is a martingale argument. Let $\mathfrak{J}_t$ be the σ-field generated by $\{X_s, 0 \leq s \leq t\}$ and all null sets, so that $\mathfrak{J}_{t+} = \mathfrak{J}_t$ for $t \geq 0$; and for any optional τ let $\mathfrak{J}_\tau$ and $\mathfrak{J}_{\tau+}$ and $\mathfrak{J}_{\tau-}$ have the usual meanings.

Lemma 2. *If* $a < b < \beta$, *then*

$$\lim_{a\uparrow b} u(a,b) = 1\,; \tag{7}$$

$$\lim_{b\downarrow a} u(a,b) = 1\,. \tag{8}$$

Proof. Let $a < b_n \uparrow b$ and consider

$$E_a\{e(\tau_b)|\mathfrak{J}(\tau_{b_n})\}\,, \qquad n \geq 1\,. \tag{9}$$

Since $b < \beta$, $u(a,b) < \infty$ and the sequence in (9) forms a martingale. As $n \uparrow \infty$, $\tau_{b_n} \uparrow \tau_b$ a.s. and $\mathfrak{J}(\tau_{b_n}) \uparrow \mathfrak{J}(\tau_{b-})$. Since $e(\tau_b) \in \mathfrak{J}(\tau_{b-})$, the limit of the martingale is a.s. equal to $e(\tau_b)$. On the other hand, the conditional probability in (9) is also equal to

$$E_a\left\{e(\tau_b)\exp\left(\int_{\tau_{b_n}}^{\tau_b} q(X(s))ds\right)|\mathfrak{J}(\tau_{b_n})\right\} = e(\tau_{b_n})u(b_n,b)\,.$$

As $n \uparrow \infty$, this must then converge to $e(\tau_b)$ a.s.; since $e(\tau_{b_n})$ converges to $e(\tau_b)$ a.s., we conclude that $u(b_n, b) \to 1$. This establishes (7).

Now let $\beta > b > a_n \downarrow a$, and consider

$$E_a\{e(\tau_b) | \mathfrak{J}(\tau_{a_n})\}, \qquad n \geq 1. \tag{10}$$

This is again a martingale. Although $a \to \tau_a$ is a.s. left continuous, not right continuous, for each fixed a we do have $\tau_{a_n} \downarrow \tau_a$ and $\mathfrak{J}(\tau_{a_n}) \downarrow \mathfrak{J}(\tau_a)$. Hence we obtain as before $u(a_n, b) \to u(a, b)$ and consequently

$$u(a, a_n) = \frac{u(a, b)}{u(a_n, b)} \to 1.$$

This establishes (8).

The next result illustrates the basic probabilistic method.

Theorem 1.[*] *The following three propositions are equivalent:*
(i) $\beta = +\infty$;
(ii) $\alpha = -\infty$;
(iii) *For every a and b, we have*

$$u(a, b)u(b, a) \leq 1. \tag{11}$$

Proof. Suppose $x(0) = b$ and let $a < b < c$. If (i) is true then $u(b, c) < \infty$ for every $c > b$. Define a sequence of successive hitting times T_n as follows (where θ denotes the usual shift operator):

$$S = \tau_a \wedge \tau_c \qquad T_o = 0, \qquad T_1 = S,$$

$$T_{2n} = T_{2n-1} + \tau_b \circ \theta_{T_{2n-1}}, \qquad T_{2n+1} = T_{2n} + S \circ \theta_{T_{2n}}, \tag{12}$$

---

[*]This odd result looks like a case of time-reversibility. Nobody has been able to authenticate this surmise. (March 2001).

for $n \geq 1$. Define also

$$N = \min\{n \geq 0 | T_{2n+1} = \tau_c\}. \tag{13}$$

It follows from $P_b\{\tau_c < \infty\} = 1$ that $0 \leq N < \infty$ a.s. For $n \geq 0$, we have

$$E_b\{e(\tau_c); N = n\} = E_b\left\{\exp\left(\sum_{k=0}^{2n}\int_{T_k}^{T_{k+1}} q(X(s))ds\right)\right\}$$

$$= E_b\{e(\tau_a); \tau_a < \tau_c\}^n$$

$$\times E_a\{e(\tau_b)\}^n E_b\{e(\tau_c); \tau_c < \tau_a\}. \tag{14}$$

Since the sum of the first term in (14) over $n \geq 0$ is equal to $u(b, c) < \infty$, the sum of the last term in (14) must converge. Thus we have

$$E_b\{e(\tau_a); \tau_a < \tau_c\}u(a, b) < 1. \tag{15}$$

Letting $c \to \infty$ we obtain (14). Hence $u(b, a) < \infty$ for every $a < b$ and so (ii) is true. Exactly the same argument shows that (ii) implies (iii) and so also (i).

We are indebted to R. Durrett for ridding the next lemma of a superfluous condition.

Lemma 3. *Given any $a \in R$ and $Q > 0$, there exists an $\varepsilon = \varepsilon(a, Q)$ such that*

$$E_a\{e^{Q\sigma_\varepsilon}\} < \infty \tag{16}$$

where

$$\sigma_\varepsilon = \inf\{t > 0 | X(t) \notin (a - \varepsilon, a + \varepsilon)\}.$$

Proof. Since X is strong Markov and has continuous paths, there is no "stable" point. This implies $P_a\{\sigma_\varepsilon \geq 1\} \to 0$ as $\varepsilon \to 0$ and so there exists ε such that

$$P_a\{\sigma_\varepsilon \geq 1\} < e^{-(Q+1)}. \tag{17}$$

Now σ_ε is a terminal time, so $x \to P_x\{\sigma_\varepsilon \geq 1\}$ is an excessive function for the process X killed at σ_ε. Hence by standard theory it is finely continuous. By for a diffusion under hypothesis (2) it is clear that fine topology coincides with the Euclidean. Thus $x \to P_x\{\sigma_\varepsilon \geq 1\}$ is in fact continuous. It now follows that we have, further decreasing ε if necessary:

$$\sup_{|x-a|<\varepsilon} P_x\{\sigma_\varepsilon \geq 1\} < e^{-(Q+1)}. \tag{18}$$

A familiar inductive argument then yields for all $n \geq 1$.

$$P_a\{\sigma_\varepsilon \geq n\} < e^{-n(Q+1)} \tag{19}$$

and (16) follows.

Lemma 4. *For any $a < \beta$ we have*

$$u(a,\beta) = \infty; \tag{20}$$

for any $b > \alpha$ we have $u(b,\alpha) = \infty$.

Proof. We will prove that if $u(a,b) < \infty$, then there exists $c > b$ such that $u(b,c) < \infty$. This implies (20) by Lemma 1, and the second assertion is proved similarly.

Let $Q = \|q\|$. Given b we choose a and d so that $a < b < d$ and

$$E_b\{e^{Q(\tau_a \wedge \tau_d)}\} < \infty. \tag{21}$$

This is possible by Lemma 3. Now let $b < c < d$; then as $c \downarrow b$ we have

$$E_b\{e(\tau_a); \tau_a < \tau_c\} \le E_b\{e^{Q(\tau_a \wedge \tau_d)}; \tau_a < \tau_c\} \to 0 \qquad (22)$$

because $P_b\{\tau_a < \tau_c\} \to 0$. Hence there exists c such that

$$E_b\{e(\tau_a); \tau_a < \tau_c\} < \frac{1}{u(a,b)}. \qquad (23)$$

This is just (15) above, and so reversing the argument there, we conclude that the sum of the first term in (14) over $n \ge 0$ must converge. Thus $u(b,c) < \infty$, as was to be shown.

To sum up:

Theorem 2. *The function* $(a,b) \to u(a,b)$ *is continuous in the region* $a \le b < \beta$ *and in the region* $\alpha < b \le a$. *Furthermore, extended continuity holds in* $a \le b \le \beta$ *and* $\alpha \le b \le a$, *except at* (β, β) *when* $\beta < \infty$, *and at* (α, α) *when* $\alpha > -\infty$.

Proof. To see that there is continuity in the extended sense at (a, β), where $a < \beta$, let $a < b_n \uparrow \beta$. Then we have by Fatou's lemma

$$\lim_{n \to \infty} u(a, b_n) \ge E_a\left\{ \lim_{n \to \infty} e(\tau_{b_n}) \right\}$$

$$= E_a\{e(\tau_\beta)\} = u(a, \beta) = \infty.$$

If $\beta < \infty$, then $u(\beta, \beta) = 1$ by definition, but $u(a, \beta) = \infty$ for all $a < \beta$; hence u is not continuous at (β, β). The case for α is similar.

3. The Connections

Now let X be the standard Brownian motion on R and q be bounded and continuous on R.

Theorem 3. *Suppose that $u(x, b) < \infty$ for some, hence all, $x < b$. Then $u(\cdot, b)$ is a solution of the Schrödinger equation:*

$$\frac{1}{2}\varphi'' + q\varphi = 0$$

in $(-\infty, b)$ satisfying the boundary condition

$$\lim_{x \to b} \varphi(x) = 1.$$

There are several proofs of this result. The simplest and latest proof was found a few days ago while I was teaching a course on Brownian motion. This uses nothing but the theorem by H. A. Schwarz on generalized second derivative and the continuity of $u(\cdot, b)$ proved in Theorem 2. It will be included in a projected set of lecture notes.* An older proof due to Varadhan and using Ito's calculus and martingales will be published elsewhere. An even older unpublished proof used Kac's method of Laplace transforms of which an incorrect version (lack of domination! see Part I, p. 11) had been communicated to me by an *ancien collègue.*

But none of these proofs will be given here partly because they constitute excellent exercises for the reader, and partly because the results have recently been established in any dimension (for a bounded open domain in lieu of $(-\infty, b)$).

These are in the process of consolidation and extension. (See Part I, References [F].)

I am indebted to Pierre van Moerbeke for suggesting the investigation in this note. The situation described in Theorem 1 for the case of Brownian motion apparently means the absence of "bound states" in physics!

---

*Part II.